Waterbody Hydrodynamic and Water Quality Modeling

An Introductory Workbook and CD-ROM on Three-Dimensional Waterbody Modeling

JOHN ERIC EDINGER

PRESS

American Society of Civil Engineers
1801 Alexander Bell Drive
Reston, Virginia 20191–4400

Abstract: This workbook provides software for a three-dimensional hydrodynamic and transport model that can be used to study the circulation in water bodies as influenced by inflows, outflows, tides, winds, salinity, and temperature. Coupled with the model is a dissolved oxygen depression model, a particulate-based nutrient and eutrophication model, and a sediment scour and deposition model. The model is set up using provided input data files. The workbook goes through the steps of setting up an example application for its bathymetry, its input data file, and specification of the desired output, followed by execution of the example. Further example applications are presented showing the setup of the model for different types of water bodies and for the different water quality models provided in the software. The results of the 30 example applications are provided in example output folders. Following the example applications are chapters on the theoretical basis and formulation of the hydrodynamic and transport model, the first-order decay relationships, the surface heat exchange relationships, the dissolved oxygen depression model, the nutrient and eutrophication model, and the sediment scour and deposition model.

Library of Congress Cataloging-in-Publication Data

Edinger, John Eric.
   Waterbody hydrodynamic and water quality modeling : an introductory workbook to
numerical three-dimensional modeling / John Eric Edinger.
      p. cm.
   "Including software, users manual, example applications, model derivations and descriptions."
   Includes bibliographical references and index.
   ISBN 0–7844–0550–6
      1. Hydrodynamics—Computer programs. 2. Water quality—Computer programs. I. Title.

TC171 .E37 2001
627'.042--dc21   2001034123

DEDICATION

This book is dedicated to my wife, Judy, and our two daughters, Frances Wilde and Susan Edinger. Frances provided advice at different stages of manuscript preparation, and assisted in handling different stages of publication. Susan arranged the Webb Institute of Naval Architecture lectures where student questions and feedback provided encouragement to proceed in developing the software and workbook for publication.

ACKNOWLEDGMENTS

Everyone on the staff at J. E. Edinger Associates Inc. has contributed to this workbook. Edward M. Buchak, my business partner since 1974, provided the encouragement to proceed and made time and resources available. Ed and I developed many of the fundamental algorithms for these types of computations in our early work on LARM (now CE-QUAL-W2) and its three-dimensional successors. Venkat Kolluru provided a computational framework for the INTROGLLVHT that is a simplified version of the full GEMSS modeling system discussed in Chapter 9. Lynn Jarrett, Rajeev Jain, and George Krallis of J. E. Edinger Associates Inc. each contributed in different ways through discussions of the text, prior modeling applications, research of background data, and reviewing the workbook and the model computations and operation.

Earlier versions of the draft text and software were tested and used by several people. Dr. John Gordon at Tennessee Technological University used the material for student class projects in his course in reservoir water quality modeling. Dr. Gordon's class included civil engineering, biology, and physics students. Questions from the students about setting up the model led to the development of the bathymetric setup routine, the input file routine, and the development of a uniform nomenclature for input and output file titles. Dr. Seok Park of Ewha University, Seoul, Republic of Korea, had students use an earlier version to perform preliminary runs of river reservoir water quality studies.

Kristin Geotchius, a graduate student in The Johns Hopkins University Department of Geography and Environmental Engineering, tested the software for initial studies of the fate of pesticides in upper Chesapeake Bay and recommended inclusion of the groundwater inflow input data routine. Dr. Haydee Salmun, advisor to Kristin Geotchius and now at Hunter College, and Dr. Jon Hubertz of Applied Coastal Modeling reviewed an earlier version of the workbook and both provided many useful suggestions for improving the text.

Special recognition goes to the late Chuck Boatman of Aura Nova, Inc. who contributed indirectly to this text through our years of mutual work on various water body water quality modeling studies and the development of higher-order water quality models. Chuck was an outstanding biochemical oceanographer and modeler who was too busy to get his important contributions in the field published. He was always on the lookout for missing processes rather than manipulation of parameters when there was disagreement between water quality model results and field data. Chuck contributed much to the development of the particulate-based model used in this workbook and the carbon-based model discussed in Chapter 13. He extended the carbon-based model to include vertical migration of dinoflagellates to describe nighttime depletion of dissolved oxygen in deeper waters and production of dissolved oxygen maxima below the water surface during the day. Chuck will be missed as a professional water quality modeler and a good friend.

ABBREVIATIONS

BOD	Biochemical oxygen demand
CBOD	Carbonaceous biochemical oxygen demand
CBOD_d	Dissolved carbonaceous biochemical oxygen demand
CBOD_p	Particulate carbonaceous biochemical oxygen demand
CO_2	Carbon dioxide
CSHE	Coefficient of surface heat exchange
DO	Dissolved oxygen
DOD	Dissolved oxygen deficit
GEMSS	Generalized Environmental Modeling System for Surfacewaters
GLLVHT	Generalized Longitudinal Lateral Vertical Hydrodynamic Transport
INTROGLLVHT	Introductory version of GLLVHT
NH_3	Ammonia
NH_4	Ammonium
NO_3	Nitrate
ON	Organic nitrogen
ON_P	Particulate organic nitrogen
OP	Organic phosphorous
OP_P	Particulate organic phosphorous
PO_4	Phosphate
RDECAY	Rate of decay
SED	Sediment scour and deposition model
SOD	Sediment oxygen demand
SPO	Spatial output file
TEQ	Equilibrium temperature of surface heat exchange
TSC	Temperature salinity arbitrary constituent model
TSO	Time series output file
WQDPM	Water quality dissolved particulate model

CONTENTS

INTRODUCTION

This workbook on hydrodynamic and water quality modeling evolved from workshops and seminars in which the participants requested introductory software for learning water body modeling techniques and problem solving. It is designed for students in environmental engineering and environmental sciences programs to supplement their courses in water quality and to provide hands-on experience in combined numerical hydrodynamic and water quality modeling. The models included in the workbook are suitable for use as class assignments and term projects, and have sufficient detail and flexibility that they can be used for many types of creative studies relevant to a wide range of water quality problems and practical applications.

The introductory models are designed to simulate steady flow situations where the inflows into a water body equal the outflows, or for stationary-state situations when tidal elevation boundary conditions are specified. The introductory models are used in practice for many purposes, including the following:

- making a rapid assessment of a water body problem

- determining sensitivity to model grid detail and to changes in bathymetry

- determining sensitivity of the model results to different parameters such as surface winds and water quality rate parameters

- determining the response to waste discharge rates, intake and discharge locations, river inflow rates, and different tidal conditions

- aiding in the design of field data sampling programs

The modeling of water quality throughout a water body can be done only to the detail to which the flow field transporting the water quality constituents is known. The flow field in a water body can most accurately be computed using a hydrodynamic and transport model to which the water quality model is coupled. The hydrodynamic and transport model presented in this workbook is the Generalized Longitudinal Lateral Vertical Hydrodynamic and Transport model (GLLVHT) described in Edinger and Buchak (1995). The GLLVHT model and four water quality routines are incorporated in the INTROGLLVHT operating system that couples the models with input data and allows one to specify the kinds of outputs desired. INTROGLLVHT is a simplified learning version of the more complete and detailed Generalized Environmental Modeling System for Surfacewaters (GEMSS), to which it is compared in Chapter 9.

The GLLVHT hydrodynamic model is driven by the inputs of inflows and outflows, boundary tides and winds, and horizontal density gradients due to stratification that results from the circulation. Background to the hydrodynamic and transport relations and organization of the INTROGLLVHT computations are presented in Chapter 1,

Section 1.1. The complexities of the flow regimes that can be generated by GLLVHT are presented in Section 1.2.

Three water quality models and a sediment model are provided for coupling to the hydrodynamic and transport model. These are the following:

- a temperature, salinity, and arbitrary first-order decay constituent model (TSC)

- a dissolved oxygen depression model (DOD)

- a eutrophication model (WQDPM)

- a sediment scour and deposition model (SED)

The water quality models allow one to study a wide range of realistic water quality problems in all types of water bodies including river reaches, lakes, reservoirs, estuaries, and coastal waters. The models allow one to start with the simplest case (TSC) to determine the effects of inflows, outflows, tide, wind, and density stratification on circulation and residence time throughout a water body. The TSC model allows one to study the distributions within a water body of the fate of first-order decay constituents, dilution of discharges, residence times, salinity distributions, and temperature distributions due to inflows, heat sources, and surface heat exchange. The DOD model allows one to study the impact of a discharge on lowering the dissolved oxygen within a water body without having to do full dissolved oxygen modeling. The WQDPM model performs complete dissolved oxygen modeling and allows one to study algal growth and eutrophication. The SED model allows one to study the sediment scour and deposition rates as influenced by bathymetry and by structures such as breakwaters. Most water body water quality modeling studies will progress from one model to the other as more complex features of a particular problem are examined. Background to the water quality models is presented in Chapter 1, Section 1.3.

The model software includes routines for setting up the water body bathymetry file, the model input data file, and the routine for executing the combined hydrodynamic and water quality model. The bathymetry routine is presented in Chapter 1, Section 1.5, along with its application to an example project that is followed through the next two chapters to execution. The input routine is presented in Chapter 2 along with its application to the same example project. These two steps allow the user to execute the example application in Chapter 3. It is important that the user work through these chapters to learn input file notation and how to apply the software before trying to set up an independent project.

Chapters 4 and 5 show the setup of the hydrodynamic and transport models for different types of water bodies including estuaries, coastal waters, and lakes. They illustrate the use of different combinations of boundary conditions, and provide a description of the circulation, salinity, and temperature structure that results.

Chapters 6, 7, and 8 present the application of the DOD, WQDPM, and SED models. Basic properties of the DOD and WQDPM models are illustrated using tank test and stream simulation results. The latter allow one to compare some of the numerical simulation results to analytical solutions for the simpler cases. Each water quality model is then applied to estuary, lake or reservoir, and coastal water projects. These example applications illustrate how to set up different types of boundary conditions and how the water quality within a water body adjusts to inflows, outflows, discharges, tides, and wind. The theoretical basis of the hydrodynamic and transport relationships and the details of the different water quality models used in INTROGLLVHT are presented in Chapters 9 through 14.

More than 30 example applications are presented in Chapters 4 through 8. The preparation of so many example applications led to the file naming and organization presented in Chapter 1, Sections 1.6 and 1.7. The example applications were all reworked during the final preparation of the workbook to test the executable form of the model codes. Presented with each example application is a set of problem exercises for further study that illustrate additional techniques and uses of the models. Not all of the problem exercises for futher study have been set up and tested.

This workbook presents a water body hydrodynamic and transport model and a set of water quality models that can be run with it. The workbook is mostly about how to use the INTROGLLVHT modeling system, how to set up the models for different types of water bodies and water body problems, and different ways of examining and interpreting the model results. The workbook allows one to learn different water body hydrodynamic and water quality modeling techniques and learn the properties of different models by operating them. The workbook is not intended to be a treatise on water body hydrodynamics or on water quality modeling. A good background to water body hydrodynamics as applied to water quality modeling can be found in Martin and McCutcheon (1999). The ASCE Engineering Mechanics Division book on Environmental Fluid Mechanics (Shen, et al. 2002) includes good basic descriptions of turbulent processes, estuarine hydrodynamics and basic mechanics and case studies of water quality modeling in lakes and reservoirs. The classic work on water quality modeling is Thomann and Mueller (1987). A more recent volume on water quality modeling is Chapra (1997). Another introductory text on water quality modeling is Lung (1993). An important document for the formulation of water quality model processes and associated rates, constants, and kinetics is EPA (1985). These six works should be consulted for more in-depth background in water body hydrodynamics and water quality modeling. A good qualitative treatise on limnology with plenty of illustrative data is Wetzel (2000). A good volume on many estuarine topics is Lauff (1967).

A NOTE ABOUT THE SOFTWARE

For help in running the software, please write a brief message of what problem seems to be occurring, and attach your _BATH.dat, _INP.dat and _WQM.dat files to an email addressed to John.Edinger@jeeai.com

System Requirements

i486™ or Pentium® processor-based personal computer
Microsoft Windows® 95, Windows® 98, Windows NT® 4.0 or Windows® 2000
20 MB of available RAM
Minimum 30 MB of available hard-disk space (16 MB are required for the program. 14
MB are required for Acrobat Reader)

Installation Instructions

1. Insert CD into CD-ROM drive. Autorun should start the installation after the CD has
   been inserted in the CD-ROM drive. If it doesn't start, then follow steps 2–4.
2. Select Run from the Start menu.
3. Type "e:\setup.exe", where "e" is the CD-ROM drive name/letter.
4. Press Enter key or click OK button.
5. Follow instructions provided by the software on your computer monitor.
6. The software will not be installed under any program group.

Uninstall Instructions

1. Insert CD into CD-ROM drive
2. Click the Next button
3. Select "Remove"
4. Follow instructions provided by the software.

Operating Instructions

Read "ExecutableProgramsandFilesinIntrogllvhtModelFolder.pdf" (located in the sub-
directory *Introgllvht Model With DMAWin* under the root directory *Book Software and
Applications*) using Acrobat Reader 4.05 for a description of the model program files and
information as how to run the program using *DMAWin*. Access the website
www.adobe.com for support as how to install the Acrobat Reader on your computer.

Technical Support

Should you have any questions or problems with the installation of this software or with
accessing the example applications, please contact the author at John.Edinger@jeeai.com
with a brief informative description of the problem along with the _bath.dat, _inp.dat and
_wqm.dat project files.

Copyright/Licensing Agreement

COPYRIGHT

LICENSE GRANT

1. INTROGLLVHT HYDRODYNAMIC AND WATER QUALITY MODELING

In this chapter, the fundamentals of the Generalized Longitudinal Lateral Vertical Hydrodynamic and Transport (GLLVHT) model are presented along with various examples of the kinds of flow regimes it can generate even for simple input conditions. Details of four water quality models are given. This is followed by a presentation of the contents of the INTROGLLVHT modeling system folder included on the CD-ROM. In addition, the use of the bathymetry setup routine included in the modeling system is presented with an example application. Finally, an important part of the INTROGLLVHT modeling system in organizing input and output file names for a project application is described.

1.1 The Hydrodynamic and Transport Model

The GLLVHT hydrodynamic and transport model, shown schematically in Figure 1-1, computes the horizontal and vertical velocity components (U,V,W), the water surface elevation (Z), and the density and constituent concentrations (ρ, C) at each model grid cell over time. The six basic relationships that are used to compute the six unknowns are as follows:

- the horizontal fluid momentum in the x-direction, used to evaluate U

- the horizontal fluid momentum in the y-direction, used to evaluate V

- the internal continuity of fluid, used to evaluate W

- the continuity of fluid integrated to the free water surface, used to evaluate Z
  the continuity of mass for n constituents, used to evaluate C_n

- an equation of state relating fluid density to temperature and salinity, $\rho(T,S)$

Fluid density enters the horizontal fluid momentum relationships as horizontal density gradients. The horizontal density gradients are important internal driving forces that produce density flows within a water body. There is feedback between the constituent relationships and the momentum relationships, because the former are used to determine the temperature and salinity distributions throughout the water body.

The external inputs (or boundary conditions) for a water body model are the inflows, the outflows, and the temperature of the inflows. For tidal boundaries, the external inputs are tidal elevation, salinity, and temperature at the boundary. Then there is surface winds, surface heat exchange, and bottom friction. For water quality modeling, there is the concentration of constituents in the inflows and at the tidal boundaries for each constituent included in the particular water quality model being used.

The GLLVHT is a time-varying, finite difference numerical model. It is set up in the INTROGLLVHT modeling system for computations on a rectangular grid with vertical layers of uniform thickness, except for the top layer, which varies in thickness spatially and temporally. As will be shown in Chapter 9, the solution technique used in the GLLVHT model is derived by numerically substituting the horizontal momentum balances into vertically integrated continuity to arrive at a relationship for time-varying surface elevations in the two horizontal directions. The surface elevations in the two horizontal directions are solved simultaneously on each time step. This is called an implicit solution, which is not limited by the computational time step. Because of the substitution of the U and V momentum relationships into vertically integrated continuity, the surface elevation Z and the U, V, and W velocity components are solved for simultaneously.

Table 1-1 presents an outline of the numerical computational code; it shows the steps to be taken, from reading the input files from the control files to using the data they supply to perform the hydrodynamic and water quality transport computations. The forcing functions evaluated in Step 9 include the spatial momentum terms, the surface wind shear, internal velocity shear between the layers, and bottom friction. The surface elevation, Z, is evaluated implicitly, and then the velocity components are computed. The velocity components and surface elevation along with the source/sink terms for the particular water quality model being used are inputs to the constituent transport relationship for evaluating constituent concentrations. The results of the computations for the time step are placed in the various output files, and the computations are repeated for the next time step.

The major conditions and limitations of the form of the GLLVHT model used in the INTROGLLVHT system, and of the INTROGLLVHT system itself, are shown in Chapter 9 by comparing it with a more complete version of the model and modeling system.

1.2 Example Flow Regimes

Examples of different types of flow regimes generated by the GLLVHT hydrodynamic model for simple water body configurations are presented and examined in this section. Examples of baroclinic flow and barotropic flow show that relatively complex circulation patterns are established even for simple boundary conditions imposed on a long rectangular channel. Another example is presented that demonstrates the velocity field the GLLVHT model produces when the Coriolis acceleration dominates. The distribution of water quality constituents within a water body can only be determined to the accuracy that the flow field is known. These examples demonstrate the need to perform detailed hydrodynamic computations, with which water quality models can then be coupled.

1.2.1 Barotropic Flows

Part of the flow field within a water body results from the surface slope through the water body. The component of flow related to the surface slope is called the barotropic flow. An example of a barotropic flow field is the result of wind shear on the surface of a long channel. The surface slope along the channel and the midchannel velocity profile are shown in Figure 1-2 for a 4 m/s surface wind. The surface wind moves currents in the surface layer in the direction of the surface wind, and the latter produces a surface slope setup in the direction the wind is blowing. The surface setup produces a bottom layer current flowing in the opposite direction. The bottom currents are due to the surface slope and move from the higher surface elevation toward the lower surface elevation. The bottom current is a barotropic flow. The surface current is approximately 2% of the surface wind speed that is typical of observed values.

1.2.2 Baroclinic Flows

The component of flow related to the horizontal density differences is called the baroclinic flow. A baroclinic or density-induced flow is seen, for example, when there is a vertical salinity profile at the mouth of the channel, with lower salinity water in the top layer and higher salinity water in the lower layer. For this example, there is no freshwater inflow at the closed end head of the channel. The velocity profile near the mouth of the channel and the salinity profiles near the mouth and the head of the channel are shown in Figure 1-3. The velocity profile shows an up channel surface and bottom flow and a midlayer outflow.

Examination of the density profiles shows that the lighter surface water at the mouth of the channel is sliding over heavier surface water at the head of the channel, and the heavier bottom water at the mouth of the channel is sliding under the lighter bottom water at the head of the channel. The density profiles also show that significant flows can develop with relatively small longitudinal density differences. The midlayer outflow from the head of the closed channel is equal to the sum of the surface and bottom inflows. Cameron and Pritchard (1965) demonstrated the existence of this type of three-layered baroclinic circulation in Baltimore Harbor.

The outcome of discharges into a closed channel with a baroclinic flow is strongly related to the flow pattern. Nutrient constituents entering the channel with surface runoff would first flow up the channel and return as a midlayered flow. This could have consequences relative to processes such as reaeration, temperature structure as related to surface heat exchange, and how much of the loading actually flows out of the channel.

Further analyses would show that the greater the differences in density between the surface and bottom at the mouth of the channel, the stronger the three-layered baroclinic flow. The largest density difference is produced when the salinity profile at the head of the channel is uniform from top to bottom at a concentration equal to the average salinity over the profile at the mouth of the channel. The latter requires a longer channel than used in the example.

1.2.3   Combined Barotropic and Baroclinic Flows

Figure 1-4 is the velocity profile resulting from a combination of the salinity profile boundary condition at the mouth of the channel and a wind of 4 m/s blowing from its head to its mouth. It illustrates how complex the flow field can become when both the barotropic and baroclinic forces are acting on a water body at the same time.

1.2.4   Effects of Coriolis Acceleration

The GLLVHT hydrodynamic and transport model includes the Coriolis acceleration resulting from the earth's rotation, which becomes important in very large water bodies. It can be illustrated by simulating a large 100 km by 100 km basin using a 5 km by 5 km grid with a 5 m/s wind blowing from south to north on the surface. The basin is at 40° north latitude. The results are given in Figure 1-5 as the velocity spiral of V versus U over depth for a 20 m and 60 m deep basin.

These results differ from the ideal analytical Ekman spiral solution in which the surface current is at an angle of 45° to the right of the surface wind. This is because the basin is not of infinite horizontal extension or depth, and because the vertical eddy viscosity is not constant but rather a function of vertical shear resulting from the vertical velocity profiles. The results in Figure 1-5 are similar to those obtained by Neumann and Pierson (1966), who showed that the ideal Ekmann result is approached as the basin deepens. In both cases, the surface current speed is approximately 3% of the wind speed, which is consistent with previous observations.

The Coriolis acceleration can deflect tidal currents and currents due to freshwater inflows in large bays and estuaries. For example, it is an important force affecting the wind- and inflow-induced circulation in the Great Lakes.

1.3 The Water Quality Models

Four water quality models are described in this section. They are numbered and identified as follows:

1. temperature, salinity, first-order decay constituent (TSC) model

2. the dissolved oxygen deficit (DOD) model

3. dissolved- and particulate-based eutrophication (WQDPM) model

4. the sediment scour and deposition (SED) model

The constituents included in each of the water quality models are summarized in Table 2-2. The constituent transport of heat (temperature) and salinity is computed regardless of the water quality model being run. They are required for the computation of density, which is important to the momentum balances and temperature-dependent kinetic reaction rates in the different water quality models.

1.3.1   The Temperature, Salinity, First-Order Decay Constituent Model

The TSC model is presented by itself so that circulation computations can be performed before the more elaborate water quality models are run. The TSC model includes a constituent that can undergo first-order decay and the surface heat exchange terms for temperature computations. Both are applied in Chapter 4 and Chapter 5. The first-order decay relationships are derived in Chapter 10, and the surface heat exchange computation used to evaluate temperature is derived in Chapter 11.

1.3.2   The Dissolved Oxygen Deficit Model

The DOD model computes how much the dissolved oxygen is changed in a water body due to inflows of organic nitrogen, ammonium, and biochemical oxygen dem. The organic nitrogen mineralizes to ammonium. The ammonium takes up dissolved oxygen as it oxidizes to nitrate. The oxygen uptake is balanced by surface reaeration.

Each oxygen uptake stage has a temperature-dependent kinetic reaction rate. The rate of surface reaeration is temperature- and wind speed-dependent. The dissolved oxygen deficit model is applied in Chapter 6 and derived in Chapter 12.

1.3.3   The Dissolved- and Particulate-Based Eutrophication Model

The WQDPM model computes the dissolved oxygen in a water body due to inflows of the dissolved forms of organic nitrogen, ammonia and ammonium, biochemical oxygen demand, organic phosphorous, phosphate, and the particulate forms of organic nitrogen, biochemical oxygen demand, and organic phosphorous. The WQDPM model includes phytoplankton and zooplankton grazing. The nitrogen and biochemical oxygen demand components undergo oxidation as in the DOD model. The phytoplankton produce oxygen during the day (photosynthesis) and take up dissolved oxygen at night (respiration), and the rates are nutrient-concentration dependent. The phytoplankton produce dissolved organic nitrogen and organic phosporous, and re-cycle carbon to biochemical oxygen demand. Zooplankton grazing produces particulate organic nitrogen and organic phosphorous, some of which can settle out.

Each of the individual reaction rates is not only temperature dependent, but also can be oxygen and nutrient-concentration limiting. The particulate constituents can settle out, and there is a release of nutrients back into the water column. The WQDPM eutrophication model is applied in Chapter 7 and derived in Chapter 13.

1.3.4   The Sediment Scour and Deposition Model

The SED model computes a scour rate and a sediment settling rate to determine a net scour or deposition rate along the bottom of the water body. The scour rate computation is performed from a semiempirical model. The rates depend on the sediment particle diameter and specific gravity, and the density of the surrounding fluid. The SED model is applied in Chapter 8 and derived in Chapter 14.

1.4 INTROGLLVHT Model Folder

The INTROGLLVHT Model folder on the CD-ROM contains routines for setting up and executing the water body model, as well as the water quality parameter default files. The execution routines are as follows:

Bathymetrysetup.exe, which is used to set up a water body bathymetry file.

INTROGLLVHT Input File.exe, which is used to set up the water body project input data and output specifications.

Generalmodel.exe, which contains the hydrodynamic and water quality models and is used to execute them from the bathymetry, input, and water quality model files specified by the user in the *ABControls.dat* file.

A side computation file called *Teq Cshe Computation.exe* is provided for the evaluation of surface heat exchange parameters from meteorologic data and is discussed in Chapter 11. The INTROGLLVHT Model folder also contains the default parameter files for each of the water quality models that are detailed in Chapter 3.

In the following sections, each of the INTROGLLVHT Model files is discussed separately. The bathymetry routine is presented in detail and used to set up the bathymetry for the first example project application.

1.5 The Bathymetry Setup Routine and Its Application

The bathymetry setup routine called Bathymetrysetup.exe aids in the preparation of the bathymetry data file. In this section, it is presented and used to set up the bathymetry for the first model application of an example estuary.

When started, the bathymetry setup routine first asks for a file name. The name should be the water body name and the number of the bathymetric file being set up for that water body. For instance, the example name is specified as Example_01. The routine then asks for the maximum size of the computational grid, IM, JM, and KM. It further asks for the size of each grid cell ΔX, ΔY, and ΔZ. The suffix BATH of the file name is produced when the skeleton data table is generated and automatically placed in the INTROGLLVHT Model folder.

An example skeleton bathymetry table is shown in Table 1-2. It has default depths set to zero at each I and J location over the active computational grid. In most applications, there are a large number of land cells in the model grid indicated by the zero depth. The zeros are replaced in the table by the water body depth at the center of each computational grid location. The table ends with a delimiter value of –999, indicated by "delim." The latter checks to see if there are depth values at each I and J location.

The spatial arrays of all the variables are dimensioned at IM = 50, JM = 50, and KM = 30, and these dimensions define the size of the maximum grid that can be used in the INTROGLLVHT model. The number of KM layers usually governs the time it takes to complete a simulation.

The smallest grid for which computations can be performed is IM = 3 by JM = 3 (one cell) and a minimum of two active layers at all points throughout the grid. Sometimes it is useful to perform a "tank test" computation on a water quality model that has one surface cell and a few vertical active layers.

1.5.1   *The Estuary Example Application Computational Grid*

For most applications, a transparent grid is placed on a bathymetric map of the water body and the depth is written in each grid cell. The computational grid for the example application estuary is given in Table 1-3 to help define the model grid coordinates. The grid extends from I = 2 to IM-1 (IM = maximum) from west to east along the x-axis, and from J = 2 to JM-1 (JM = maximum) from south to north along the y-axis.

The cells for J = 1 from I = 1 to IM and J = JM from I = 1 to IM and the cells for I = 1 for J = 1 to JM and I = IM for J = 1 to JM are dummy cells. Dummy cells are not used in the model computations or in defining the model geometry. Each cell has dimensions of Δx by Δy. The depth of the water body at a particular I,J location is identified in each cell. The cells outside the water body have a depth of zero.

The thickness of the model layers is Δz. The top cell of the vertical model grid is at K = 2, and the bottom cell is identified as K0(I,J). The latter is computed from the depths by the GLLVHT model for each I,J surface cell. Its value is K0(I,J) = INT(depth(I,J)/Δz)+1, and it is evaluated only where the depth is not zero. The KM in the vertical should be set at:

$$KM = INT(Maxdepth/\Delta z)+4 \tag{1.1}$$

The KM can be set larger than this value. The computations in GLLVHT are performed only where K0(I,J) is not zero and only to its level at each I,J location.

1.5.2 *Application of Bathymetrysetup.exe Routine*

The Bathymetrysetup.exe routine given in the INTROGLLVHT MODEL folder is used to set up the model grid. When it is run, it will ask for the data given in Table 1-3 extending from the name of the water body (given here as Estuary) through to the Δz (or dz) value. It will then generate a file with the name Estuary_BATH.dat and with default values of zero at each I,J location. This file is then opened and the nonzero depths are substituted. The nonzero depths for the estuary example application are obtained from the model grid data given in Table 1-3. The model grid data shows zeros internally at I = 15 to I = 16 at J = 8, which represents an island. Islands can be as small as a single cell. The resulting bathymetry data file for the example application is shown in Table 1-4.

It is recommended that as a learning exercise the INTROGLLVHT user go through the steps of using Bathymetrysetup.exe by generating the Estuary_BATH.dat file in the INTROGLLVHT Model folder. This will provide the bathymetry file for this project that will be run as an example demonstration in Chapter 3. The generated Estuary_BATH.dat file can be checked for the example application by comparing it with the bathymetry file shown in the Applications folder named EST_TSC_01.

1.6 The Input File Routine

An input file setup routine called INTROGLLVHT Input File.exe aids in the preparation of the input data file. The input file setup routine will ask sequentially for the input data, such as inflows, outflows, tidal boundaries, and their locations, needed to drive the model. The definitions of terms in the input file routine and its application are presented in Chapter 2.

The input file setup routine first asks for a project name. It is recommended that the project name include the following three components: the name of the water body being examined, the water quality model being run, and the case being run. For the example project that is set up in Chapter 2 using the input file routine, Est_TSC_01 indicates that the water body is Est, the water quality model being run is the TSC model, and the setup is for study case 01 of that project. This name will then automatically become the prefix to all other input and output files for that project and case. Note that the name length should not exceed 40 characters.

The input file routine will generate the names and file suffixes for all the different types of files read as input data to the model and store the output from the simulation. The suffixes are listed in Table 1-5 to indicate the different types of input data files read by the model and the types of output resulting from a simulation.

The _BATH.dat file is generated by the bathymetry setup routine. The _INP.dat file, the skeleton water quality model file _WQM.dat, and the _Con.dat file are generated by the input file routine. The bathymetry, input, and water quality model file names go into the ABControlfiles.dat file. Their complete names are stored in the project _Con.dat file for transfer to the ABControlfiles.dat file. Notes on the project application can be made in the _Con.dat file, but these notes will not transfer to the ABControlfiles.dat file. The remaining files are generated as output when the INTROGLLVHT model is run.

The input file routine generates a skeleton (blank) file for the water quality parameters (_WQM.dat). The default parameters for the specified water quality model (the TSC, DOD, WQDPM, and SED models) are placed in this skeleton table and set to any desired values. The TSC water quality model table is always blank, because all of the parameters required to run it are specified under external parameters within the input data table. The default parameters in the DOD, WQDPM, and SED models are discussed in detail in Chapter 3, Section 3.3 and in various case applications in Chapter 4 through Chapter 8.

1.7 The Generalmodel.exe Routine

After the file names for _BATH.DAT, _INP.DAT, and _WQM.DAT of the project being run are placed into the ABControlfiles.dat file, the GLLVHT model is ready to be run. This is done using the Generalmodel.exe routine. The running of an example project and the types of problems that might arise when executing the model are presented in Chapter 3.

1.8 Example Applications

Example applications are presented for different types of water bodies and with different water quality models in Chapter 4 through Chapter 8. The bathymetry, input, and water quality model is given for each example. The examples illustrate different features of water body modeling that can be performed with the INTROGLLVHT Model system. It is intended that the user work through each of the example applications by using the routines given in the INTROGLLVHT Model folder to set up the bathymetry, input, and water quality default parameter files from the example data.

The input files needed to run a project are kept in the INTROGLLVHT Model folder while that project is being executed and studied. The output files from the project execution also appear in the same folder. The number of input and output files for a particular project can get quite large, and it can become confusing to have the files for more than one project, or even different cases for the same project, held in the INTROGLLVHT Model folder.

To keep individual project files together, an *Example Applications* folder is provided. This folder has a set of subfolders, one for each workbook example application. Each of the subfolders contains the files for that application, which are listed in Table 1-5.

Additional application folders and subfolders can be set up by the user to save the results obtained from running the suggested additional study applications and for completely new user projects. The project files are not automatically transferred from the INTROGLLVHT Model folder to the project subfolder. They need to be copied specifically from the INTROGLLVHT Model to an applications folder. Similarly, if an existing project application is being modified slightly to develop a new project, the files of the former need to be transferred into the INTROGLLVHT Model folder for modification and execution.

1.9 The DMAWin.exe

As an alternative to the file cut and paste procedure presented in Section 1.8, one will eventually want to learn to use the DMAWin.exe or the direct modeling access routine. It allows using a project folder containing an ABControlfiles.dat file along with associated project bathymetry, input and water quality files, run the model from those files in the folder, and have the output files returned to that folder.

When using DMAWin, none of the files within the project folder should be a read only file. The use of DMAWin.exe is explained further in the file named "Programs and files Included in the INTROGLLVHT Model" file included in the INTROGLLVHT Model folder.

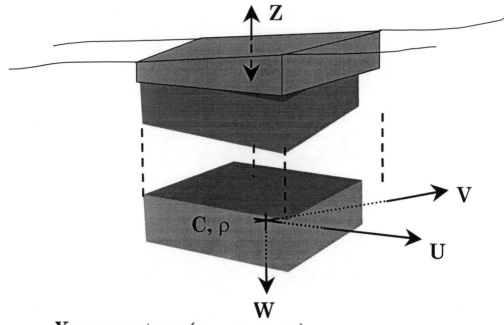

U	-	X-momentum (z, ρ, u, v, w)
V	-	Y-momentum (z, ρ, u, v, w)
W	-	Continuity (v,w)
Z	-	Vertically Integrated Continuity (v,w)
C	-	Constituent Transport (u,v,w)
ρ	-	Equation of State (C)

Figure 1-1. Definition diagram for the Generalized Longitudinal Lateral Vertical Hydrodynamic and Transport (GLLVHT) model configuration and basic relationships.

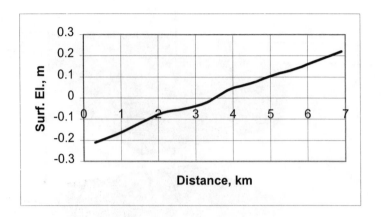

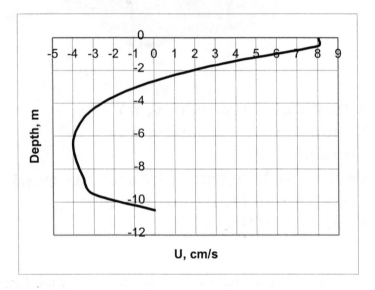

Figure 1-2. Velocity profile at the mid length of a channel produced by a surface wind illustrating a barotropic flow resulting from a surface slope.

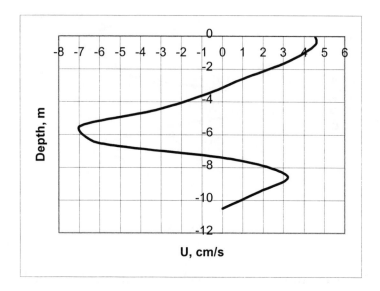

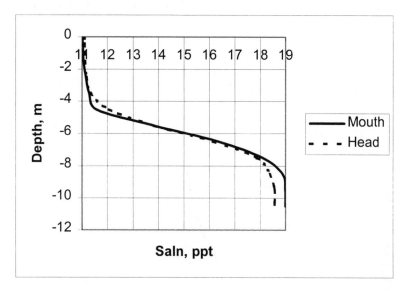

Figure 1-3. Velocity profile near the mouth of a closed channel produced by a salinity profile at the mouth of the channel. The figure also shows the resulting salinity profile at the head of the channel.

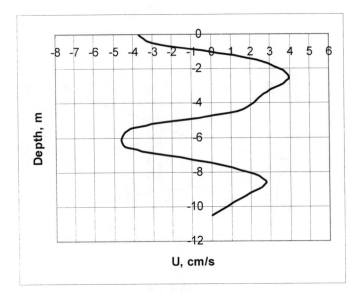

Figure 1-4. Velocity profile near the mouth of the channel for a combined boundary salinity profile at the mouth of the channel and a 4 m/s wind blowing from the head to the mouth of the channel.

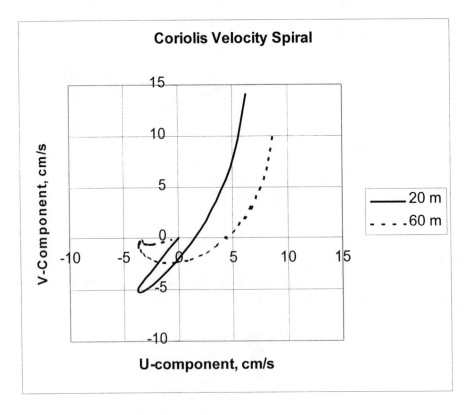

Figure 1-5. Results of simulating a large basin with a south to north surface wind of 5 m/s to demonstrate the effects of Coriolis acceleration at 20 m and 60 m depth.

Table 1-1. Steps in Generalized Longitudinal Lateral Vertical Hydrodynamic Transport (GLLVHT) hydrodynamic and water quality modeling computations.

1.   Read ABControlfiles.dat

2.   Read bathymetry data file

3.   Initialize computations (set up computational grid)

4.   Read input data file

5.   Read water quality model data file

6.   Begin time loop

7.   Evaluate inflows, outflows, elevation, and boundary conditions from input data file

8.   Evaluate water quality reaction source/sink terms for particular water quality model

9.   Compute forcing functions in each horizontal direction

10.   Compute water surface elevation implicitly from previous (ρ, Z, U,V,W)

11.   Compute U and V from barotropic and baroclinic slopes, and forcing functions

12.   Compute constituent transport from (Z,U,V,W) and source/sink terms

13.   Write out results to different files at selected intervals

14.   Continue time loop

Table 1-2. Example of default bathymetry table generated by bathymetrysetup.exe routine.

```
Example_01
         12    IM
          9    JM
          6    KM
   200.0000    DX
   200.0000    DY
   1.000000    DZ
          2    3    4    5    6    7    8    9    10   11
     8  0.0  0.0  0.0  0.0  0.0  0.0  0.0  0.0  0.0  0.0
     7  0.0  0.0  0.0  0.0  0.0  0.0  0.0  0.0  0.0  0.0
     6  0.0  0.0  0.0  0.0  0.0  0.0  0.0  0.0  0.0  0.0
     5  0.0  0.0  0.0  0.0  0.0  0.0  0.0  0.0  0.0  0.0
     4  0.0  0.0  0.0  0.0  0.0  0.0  0.0  0.0  0.0  0.0
     3  0.0  0.0  0.0  0.0  0.0  0.0  0.0  0.0  0.0  0.0
     2  0.0  0.0  0.0  0.0  0.0  0.0  0.0  0.0  0.0  0.0
       -999    geomdelim
```

Table 1-3. Estuarine bathymetric grid.

```
Estuary
 21 im
 11 jm
 13 km
500 dx
500 dy
  1 dz
```

J \ I	2	3	4	5	6	7	8	9	10	11	12	13	14	15	16	17	18	19	20
10	0	0	0	0	0	0	0	0	0	0	0	0	0	0	3	3	0	0	0
9	0	0	0	0	0	0	0	0	0	0	0	0	0	3	3	3	0	0	0
8	0	0	0	0	0	0	0	0	0	0	0	0	4	4	4	3	3	0	0
7	0	0	0	0	0	0	0	0	0	0	0	4	4	5	5	4	4	0	0
6	0	0	0	0	0	0	0	0	0	3	4	5	5	0	0	5	4	4	0
5	0	0	0	0	0	0	0	3	3	4	4	5	6	7	7	7	5	4	4
4	4	4	5	5	6	6	6	6	7	7	7	8	8	9	9	9	10	10	10
3	0	3	3	4	4	5	5	5	6	6	0	0	5	8	8	9	9	10	10
2	0	0	0	0	0	4	4	0	0	0	0	0	0	0	4	5	5	4	0

I

Table 1-4. Estuary bathymetric file, Estuary_BATH.dat, after using BATHYMETRYSETUP.EXE and filling in nonzero depths from Table 1-3.

```
Estuary
21   IM
11   jm
13   km
500  dx
500  dy
1  dz
     2  3  4  5  6  7  8  9 10 11 12 13 14 15 16 17 18 19 20
10   0  0  0  0  0  0  0  0  0  0  0  0  0  0  3  3  0  0  0
9    0  0  0  0  0  0  0  0  0  0  0  0  0  3  3  3  0  0  0
8    0  0  0  0  0  0  0  0  0  0  0  4  4  4  3  3  0  0
7    0  0  0  0  0  0  0  0  0  0  0  4  4  5  5  4  0  0
6    0  0  0  0  0  0  0  0  3  4  5  5  0  0  5  4  0  0
5    0  0  0  0  0  0  3  3  4  4  5  6  7  7  7  5  4  4
4    4  4  5  5  6  6  6  6  7  7  7  8  8  9  9  9 10 10 10
3    0  3  3  4  4  5  5  5  6  6  0  0  5  8  8  9  9 10 10
2    0  0  0  0  0  4  4  0  0  0  0  0  0  0  4  5  5  0  0
-999   geomdelim
```

Table 1-5. Summary of file suffixes for input and output files produced by the INTROGLLVHT INPUT FILE.EXE routine.

File Suffix	File Contents
_BATH.dat	Generated by bathymetry input routine
_INP.dat	Input data file
_WQM.dat	Water quality parameter file
_SPO.dat	Spatial output results file
_TSO.dat	Time series output file
_PLT_SUR.dat	Plotting data for water body surface
_PLT_BOT.dat	Plotting data for water body bottom
_PLT_PO1.dat	Plotting data for the first profile specified
_PLT_PO2.dat	Plotting data for the second profile specified
_BATH_PLT.dat	Plotting data for bathymetry
_Con.dat	Stores files required in control file AbControlfiles.dat

2. THE INPUT DATA FILE

The most complex file needed to run the INTROGLLVHT model is the input data file, suffixed as _INP.dat. It specifies the locations and flow rates of inflows and discharges into the water body, the constituent concentrations of the inflows, the outflows and withdrawal locations and rates, and the tidal boundary elevation and constituent concentrations at the boundary. In addition, it contains all of the information necessary to drive the model hydrodynamics for the particular project and water body being examined, as well as specifies the details of what will be contained in the output files. For a project simulation to run correctly, it is necessary to become familiar with the parameters that must be specified in the input data file and how they should be specified.

In this chapter, the input data file is first described in general terms and then in the form required for an example project application. The use of the INTROGLLVHT Input File.exe routine is then explained and used for the project application.

2.1 Description of Input Data File

The model input data file is a form that needs to be filled out for each project application. An idealized version of the form is shown in Table 2-1. The lines and symbols preceded by a dollar sign ($) describe what needs to be entered in the data boxes. The data in the latter are identified in *italics*. Note that the $ descriptor lines and symbols are a permanent feature of the table and should not be changed. Following is a detailed, line by line discussion of each parameter and section of the model input data file.

The first line on the form is the project name that will be printed out as the prefix to all project input and output files except for the bathymetry file. It can contain any information about the water body and the conditions for which the simulations are being run. It is useful to include in the project name the water body name, identification of the water quality model being used, and the case number for that project. These should be connected by an underscore (_) so that all of the information in the project name appears in the output file names.

2.1.1   *Water Quality Model*

Each set of input data is linked to a particular water quality model. The nwqm specifies which water quality model is being used. Table 2-2 lists the nwqm for the different water quality models.

2.1.2   *Inflow Conditions*

The ninflows is the number of inflows into the water body. Inflows are river inflows and facility discharges into the water body. The ninflows should be set to zero if there are no inflows.

Three lines of data are required for each inflow:

The first line specifies the inflow rate and location as: qinflow in m^3/s, and the iinflow, jinflow, and kinflow.

The second line specifies if the inflow to the water body is connected to an intake withdrawing from the water body. The intake is set to 0 for no connection and 1 for a connection. The location of the intake is specified as iintake, jintake, and kintake. If intake is set to zero, then default values of zero should be used for the intake location coordinates.

The third line specifies the concentrations of constituents and the temperature in the inflow. For a facility discharge connected to an intake, only the increase in concentration and temperature between the intake and discharge should be specified. The number of constituents for which concentrations need to be specified and the order in which they are specified depends on the water quality model being used. These are summarized in Table 2-2.

No blank lines should be left between the descriptor lines and the data, and no blank lines should be left between the data for the individual inflows. If there are no inflows to the water body (ninflows = 0), there are no data lines, and no blank lines should be left.

2.1.3   Outflow Conditions

The noutflows is the number of outflows or withdrawals from a water body. Outflows are releases at dams and withdrawals by a facility. If a facility discharge is connected to an intake, then there should be a withdrawal at the same flow rate and location.

There is one line of data for each outflow or withdrawal. It contains the outflow rate qoutflow in m^3/s, and its location ioutflow, joutflow, and koutflow.

2.1.4   Elevation Boundary Conditions

The nelevation is the number of elevation boundaries. Usually there is one elevation boundary for an estuarine situation, two elevation boundaries for a canal, and three or more elevation boundaries for a coastal water situation. The kts specifies the k of the active surface layer and has a default value of 2. There are some cases where kts is set at a value larger than 2.

The following lines of data are required for each elevation boundary:

The first line of data specifies the location of the ends of the elevation boundary. An elevation boundary is located at the ends of the model grid and is either parallel to the I-axis or J-axis. Its location and orientation is specified by iewest, ieeast, jesouth, and jenorth. For an elevation boundary parallel to the I-axis, iewest is the western end, ieeast is the eastern end, and jesouth and jenorth are both either set to 2 or JM-1. For an elevation boundary parallel to the J-axis, jesouth is the southern end and jenorth is the northern end, and iewest and ieeast are both either set to 2 or IM-1.

The second line of data specifies the mean elevation height *zmean* in meters, the tidal amplitude *zamp* in meters, a *time lag* in hours relative to one of the other tidal boundaries, and the tidal period *tideper* in hours. In most cases, tides are semidiurnal with a period of 12.45 hours. The zmean should always be set equal to zamp so that the water surface does not fall below the top of the model computational gird. If there is more than one tidal boundary with different zamps, then zmean should be set equal to the largest zamp. With two or more tidal boundaries, zmean should be the same at both boundaries unless the tidal data indicates there is a difference in mean tidal elevation between the two locations. If there is a boundary with a fixed elevation but no tide, then zamp is set to zero, and tideper is set to a default value of 0.0.

The next lines are constituent concentrations at the boundary, which are required for all the constituents in the water quality model and extend from the surface at k = 2 to the bottom at k = KM-2. For each line in this set of data, the first number is the vertical k-level, and then listed across are the constituent concentrations for the particular water quality model being used. The number of constituents depends on the water quality model being used. The constituents are listed from left to right in the order indicated in Table 2-2.

There should be no blank lines between data sets, and no blank lines if there are no elevation boundaries.

2.1.5   Initialize Profiles

The water quality parameters can be initialized. One has the option of initializing the water quality profiles at the beginning of a simulation, or starting at zero initial constituent values. If ninitial is set to zero, then there will be no initialization of profiles. It is set to 1 to indicate initial profiles are being specified.

If one chooses to initialize profiles, then the next set of lines are constituent concentrations, which are required for all the constituents in the water quality model and extend from the surface at k = 2 to the bottom at k = KM-2. For each line in this set of data, the first number is the vertical k-level, and then listed across are the constituent concentrations for the particular water quality model being used.

There should be no blank lines between data sets, and no blank lines if there is no initialization.

2.1.6   External Parameters

The external parameters are the Chezy coefficient of bottom friction ($m^{1/2}$/s), the wind speed component parallel to the x-axis Wx (m/s), the wind speed component parallel to the y-axis Wy (m/s), the coefficient of surface heat exchange CSHE (Watts/m^2/°C), the equilibrium temperature of surface heat exchange Teq (C), the rate of decay Rdecay (per day) of an arbitrary first-order decay constituent, and the latitude of the water body Lat, in degrees as a decimal.

It is always necessary to specify a Chezy coefficient. Bottom friction varies inversely with the Chezy coefficient, which can range from as low as 17 $m^{1/2}$/s for high bottom friction to as high as 70 $m^{1/2}$/s for low bottom friction.

The wind speed components, Wx and Wy, determine the wind shear in each direction across the surface of the water body. The speed and direction of the wind across the surface of the water body is controlled by the magnitude of the wind speed components. The wind speed averaged over a few days typically ranges from 1.5 m/s to 3.5 m/s.

The coefficient of surface heat exchange and the equilibrium temperature, CSHE and Teq, are derived in Chapter 11. The CSHE varies from 15 Watts/m^2/°C for periods of low temperature to 35 Watts/m^2/°C for periods of high temperature.

The arbitrary first-order decay constituent can be used to represent any number of constituents, including coliform die off, chlorine decay, simple toxic substances decay, and radionuclide decay. Examples of rate constants for a number of constituents are given in Chapter 10. If the Rdecay is set to zero, the arbitrary constituent can be used to represent a tracer dye to determine the dilution and shape of a mixing zone area, and to determine recirculation between a facility discharge and intake.

The latitude is required for the Coriolis acceleration computation. It is given in decimal degrees and is positive in the northern hemisphere and negative in the southern hemisphere.

2.1.7   Output Profiles

Vertical profiles of velocity components and for selected constituent concentrations can be printed out for a slice parallel to the I-axis or parallel to the J-axis. The nprofiles is set to the number of such slices that are to be printed out. If there are to be none, then nprofiles should be set equal to zero. There are four lines of data for each profile slice that is to be printed out.

The first line identifies the location and orientation of the slice. A slice is either parallel to the I-axis or J-axis. Its location and orientation is specified by ipwest, ipeast, jpsouth, and jpnorth. For a slice parallel to the I-axis, ipwest is the western end, ipeast is the eastern end, and jpsouth and jpnorth are both set equal to the value on the J-axis along which the slice goes. For a slice parallel to the J-axis, jpsouth is the southern end, jpnorth is the northern end, and ipwest and ipeast are both set equal to the value on the I-axis along which the slice goes.

The second line identifies which velocity components are to be printed out, if any. The u-velocity component is parallel to the I-axis, and the v-velocity component is parallel to the J-axis. If no velocity components are to be printed out, then zero should be set for each u-velocity, v-velocity, and w-velocity component. If velocity components are to be printed out, then unity should be set for the particular component desired. If the slice is parallel to the I-axis, then either the u-velocity or the v-velocity and w-velocity components can be printed out. If the slice is parallel to the J-axis, then either the u-velocity or the v-velocity and w-velocity components can be printed out.

The third line specifies the number of constituents, nconstits, for which profile slices are to be printed out. If no constituent profile slices are to be printed out, then nconstits should be set equal to zero.

The fourth line identifies which constituents profiles are to be printed out. In each water quality model, a number as shown in Table 2-2 identifies each constituent and that number is used here.

The four lines of data are repeated for each slice. There should be no blank lines between data sets, and no blank lines if there are no slices.

2.1.8   Output Surfaces

The surface distributions of velocity components and selected constituents can be printed out. If no surface distributions are required, then set nsurfaces to zero. For nsurfaces set to 1, the surface values will be printed out. For nsurfaces set to 2, the surface and bottom values will be printed out. There are three lines of data required for the surface printout.

The first line specifies which velocity components to print out. Either or both the u-velocity and the v-velocity components can be printed out by specifying a 0 or 1 for each.

The second line specifies the number of constituents, nconstits, for which surface distributions are desired.

The third line identifies which constituents surface distributions are to be printed out for. In each water quality model, a number as shown in Table 2-2 identifies each constituent and that number is used here. When using the sediment model, nwqm = 4, the instantaneous sedimentation/scour rate (irate) can be printed out by specifying the constituent number 4, and the long-term average sedimentation/scour rate (sedrate) can be printed out by specifying the constituent number 5.

2.1.9   Output Time Series

The time series of different constituents at different locations can be printed out to a time series file. The ntimser identifies the number of constituent/locations for which time series will be printed out. If ntimser is set to zero, there is no time series print out.

There are ntimser lines of data specifying which constituent, nconts, at what location, iconst, jconst, or konst. The nconsts identifies a particular constituent in a particular water quality model as given in Table 2-2. The time series of algal growth rate, water surface elevations, and velocity components can also be specified at any location using the constituent numbers indicated in Table 2-3.

2.1.10 Simulation Time Conditions

The first line in the simulation time conditions specifies the maximum computational time step, dtm, in seconds and the length of the simulation, tmend, in hours.

The second line specifies the frequency of the profile and surface outputs, tmeout, in hours, and the frequency of the time series output, tmeserout, in hours. If only one set of profile and surface plots are desired at the end of the simulation, then tmeout should be set equal to tmend. For a tidal case, the printouts are only for the last two tidal cycles of the simulation at a frequency of tmeout.

The dtm is limited by the Torrence condition, which states that there can be no more flow out of a cell then the volume of the cell over computational time step. Stated another way, the time step, dt, must be less than the minimum of either $\Delta x/U$, $\Delta y/V$, or $\Delta z/W$. The model will compute and display on the screen a limiting time step on each iteration. If the dtm is set too high, the limiting time step will keep decreasing until a lower value is found, or it will go to a lower value, in which case dtm should be set at a lower value.

2.1.11 Internal Boundary Locations

Internal boundaries can be set between the faces of model cells extending down from any k-layer to a lower k-level. The internal boundaries can be used to represent underflow curtain walls for selective withdrawal control from stratified water bodies, jetties extending out into a water body, and simple harbors. The internal boundaries apply at the eastward or northward face of a cell through which the U or V velocity component for that cell is computed.

The nintbnd identifies the total number of interior boundaries to be set up. If there are no interior boundaries, then nintbnd should be set equal to zero.

The next line of data identifies the location of each interior boundary. A boundary is either parallel to the I-axis or J-axis. Its location and orientation is specified by ibwest, ibeast, jbsouth, and jbnorth. For a boundary parallel to the I-axis, ibwest is the western end, ibeast is the eastern_end, and jbsouth and_jbnorth are both set equal to the value of the J-axis where the slice starts.

For a boundary parallel to the J-axis, jbsouth is the southern end and jbnorth is the northern end, and ibwest and ibeast are both set equal to the value of the I-axis along which the boundary runs. The top of the internal boundary is set at ktop and the bottom of the internal boundary extends to kbottom, which can be set equal to kmax if there is to be no underflow.

2.1.12 Constituent Average Values

It is often useful, particularly in tidal situations, to have profile and surface averages of velocity and constituent components. The section on constituent averages is designed to provide these results for each of the chosen profile and surface averages.

If no constituent averages are to be printed out, then nconarv should be set to zero. If constituent averages are desired, then nconarv should be set to the number of constituents desired. The desired constituent in each water quality model is indicated on the next line by a number as given in Table 2-2 and is used here. The time-averaged values of the velocity components specified for the output profile slices and surfaces are automatically printed out.

The constituent average value is also used to indicate that residence times at different locations in a water body should be printed out. Its use for this purpose will be illustrated by an example application in Chapter 7.

2.1.13 Groundwater Inflows

The ngrndwtr specifies if a groundwater inflow is included in the model. If it is, then ngrndwtr is set to 1; if not, then 0.

If there is a groundwater inflow, then the total groundwater inflow to the water body, qgrndwtr, is specified, and the k-levels over which it extends, kgrndU and kgrndL, are specified. Specified on the next line are the water quality constituent values of the groundwater inflow.

2.2 Application Input Data File

An application input file is first set up for a project application that uses almost every feature of the input data file except interior boundaries. The input data file for the first estuary project application, *Est_TSC_01_INP.dat*, is shown in Table 2-4. The project is set up for an estuary with salinity stratification at the tidal boundary, a river inflow, and a facility with a discharge and intake. The model grid for the estuary is given in Table 1-3 and its data file is given in Table 1-4. The facility discharge is dyed to determine recirculation and the size of its mixing zone. Reading down the table:

The project title is Est_TSC_01.

The nwqm = 1, indicating that the TSC model is being used.

The ninflows = 2, indicating that there are two inflows to the water body.

The first inflow is the river inflow at an inflow rate of qinflow = 5 m^3/s located at iinflow = 2, jinflow = 4, and kinflow = 4. This inflow is not connected to an intake, and has no constituent concentrations.

The second inflow is the facility discharge and has an inflow rate of qinflow = 1.0 m^3/s located at iinflow = 10, jinflow = 3, and kinflow = 2. The facility is connected to an intake within the estuary indicated by setting intake = 1 and located at iintake = 8, jintake = 2, and kintake = 2. For the discharge, there is no temperature or salinity increase between the intake and discharge, but a dye is injected that maintains the dye concentration increase at 100 mcg/l across the facility. With recirculation taking place, the concentration at the intake is automatically added to the increase across the facility to give the discharge concentration.

There is one outflow, or withdrawal, from the estuary, noutflows = 1, which is the facility intake or withdrawal at a flow rate of qoutflow = 1.0 located at ioutflow = 8, joutflow = 2, and koutflow = 2.

The estuary has one elevation boundary, nelevation = 1. The elevation boundary is located at iewest = ieeast = 20 (IM-1), extending from jesouth = 3 to jenorth = 5. It has a mean elevation of zmean = 0.4 m, an amplitude to zamp = 0.4 m, no time lag, and a tidal period of tideper = 12.45 hr. At the boundary the temperature profile has been set to zero. The salinity profile is 10 ppt for k = 2 to k = 4, is stepped up from k = 5 to k = 7 to represent the halocline at the boundary, and is 20 ppt for k = 8 to k = km-2. There is no dye concentration at the boundary.

The water quality profiles are initialized, ninitial = 1, from k = 2 to k = 11 at zero for temperature, 0 ppt for salinity, and zero for dye.

The external parameters are set with Chezy = 35 $m^{1/2}$/s, Wx = 0, Wy = 0, CSHE = 25 W/m^2, Teq = 0, and Rdecay = 0. With zero inflow temperatures, zero temperatures at the boundary, and zero initialized temperature, there will be no surface heat exchange with Teq = 0. Surface heat exchange could also be turned off by setting CSHE = 0. The latitude of the estuary is 30.3° north.

One profile slice, nprofiles = 1, will be in the spatial output file. It will extend from ipwest = 2 to ipeast = 20 along jpsouth = jpnorth = 4. The u-velocity = 1 component will be printed out, but v-velocity = 0 and w-velocity = 0 will not be printed out. For this profile slice two constituents, nconstit = 2, will be printed out; they are I_Const(1) = 2 (salinity) and I_Const(2) = 3 (dye).

Surface values will be printed out, nsurfaces = 2, for both the estuary surface and bottom. Printouts are for two constituents, nconstits = 2. The u-velocity component is to be printed out, u-velocity = 1, but the v-velocity component is not, v-velocity = 0. The two constituents to be printed out are I_Const(1) = 2 (salinity) and I_Const(1) = 3 (dye).

There will be time series printed out in the time series file for three constituent locations, ntimser = 3. The first and second ones are for dye, nconst = 3, located at iconst = 8, jconst = 2, kconst = 2, and iconst = 10, jcons = 4, and kconst = 2, respectively. The third one is salinity, nconst = 2, located at iconst = 2, jconst = 4, and kconst = 5.

For simulation time conditions, the maximum time step is dtm = 120 sec. The end of the simulation is set for tmend = 480 hr. The time interval of the profile, surface, and plot printouts is tmeout = 1.556 hr, which is one eighth of a tidal cycle with a 12.45-hour period. The time interval of the time series printout is tmeserout = 1.0 hr.

For the first estuary project, there are no internal boundaries, nintbnd = 0.

For constituent averages, nconarv = 2 for two constituents, which are indicated on the next line to be constituent 2 and 3 or salinity and dye.

For the first estuary project, there is no groundwater inflow.

2.3 Applying the INTROGLLVHT Input File.exe Routine

The INTROGLLVHT Input File.exe routine can be used to set up the Est_TSC_01_INP.dat file in the INTROGLLVHT folder for running the project. The INTROGLLVHT Input File.exe routine will ask for the different input data as it proceeds. Use the data in Table 2-4 as input, and have the name of the bathymetry data and the KM from that data available. For the estuary project, these are from Table 1-4, and are Estuary and KM = 13.

The input file setup routine will only ask for the data required. For example:

If two inflows are specified, it will only ask for the data for these two inflows.

For the particular water quality model specified, it will only list the water quality constituents that need values specified.

When doing profiles at tidal boundaries, it will automatically indicate the number of the tidal boundary and the **K** value for which the constituent data is to be provided and will only ask for data specifically through the **KM-2** level.

When doing initialization profiles, it will automatically indicate the **K** value for which the constituent data is to be provided and will only ask for data specifically through the **KM-2** level.

When doing profiles at tidal boundaries and initialization profiles, it will automatically ensure that only the required constituent data is provided on each line.

Through the remaining routines, it will keep track to see if the required number of data points are inserted.

If errors are made when entering the data, it is best to keep going to complete the routine, and to then go back and correct the input data file after it is placed into the

INTROGLLVHT Model folder. Check the generated input data file, named EST_TSC_01_INP.dat, against Table 2-4.

Running the INTROGLLVHT Input File.exe routine will automatically generate the EST_TSC_01_Con.dat file and the skeleton EST_TSC_01_WQM.dat file.

Table 2-1. Schematic of input file.

```
Project Name
$1.nwqm        nwqm
$2.Inflow Conditions
$ninflows      ninflows
$qinflow,iinflow,jinflow,kinflow
$intake,iintake,jintake,kintake
$temp,saln,const,bod_d,on_d,nh3,op_d,po4,phyt,bod_p,on_p,op_p
```

qinflow	iinflow	jinflow	kinflow
intake	iintake	jintake	kintake

temp	saln	const	cbod	on	nh3	do(d)	no3	op	po4	phyt	bod_p	on_p	op_p

```
$3.Outflow Conditions
$noutflows     noutflows
$qoutflow,ioutflow,joutflow,koutflow
```

qoutflow	ioutflow	joutflow	koutflow

```
$4.Elevation Boundary Conditions
$nelevations,kts      nelevation   2
$iewest,ieeast,jesouth,jenorth
$zmean,zamp,tmelag,tideper
$k,temp,saln,const,bod_d,on_d,nh3,op_d,po4,phyt,bod_p,on_p,op_p
```

iewest	ieeast	jesouth	jenorth
zmean	zamp	timelag	tideper

	temp	saln	const	cbod	on	nh3	do(d)	no3	op	po4	phyt	bod_p	on_p	op_p
2	temp	saln	const	cbod	on	nh3	do(d)	no3	op	po4	phyt	bod_p	on_p	op_p
3	temp	saln	const	cbod	on	nh3	do(d)	no3	op	po4	phyt	bod_p	on_p	op_p
4	temp	saln	const	cbod	on	nh3	do(d)	no3	op	po4	phyt	bod_p	on_p	op_p
km-2	temp	saln	const	cbod	on	nh3	do(d)	no3	op	po4	phyt	bod_p	on_p	op_p

```
$5.Initialize Water Quality Profiles
$ninitial      ninitial
$k,temp,saln,const,bod_d,on_d,nh3,op_d,po4,phyt,bod_p,on_p,op_p
```

	temp	saln	const	cbod	on	nh3	do(d)	no3	op	po4	phyt	bod_p	on_p	op_p
2	temp	saln	const	cbod	on	nh3	dc(d)	no3	op	po4	phyt	bod_p	on_p	op_p
3	temp	saln	const	cbod	on	nh3	do(d)	no3	op	po4	phyt	bod_p	on_p	op_p
4	temp	saln	const	cbod	on	nh3	do(d)	no3	op	po4	phyt	bod_p	on_p	op_p
km-2	temp	saln	const	cbod	on	nh3	do(d)	no3	op	po4	phyt	bod_p	on_p	op_p

```
$6.External Parameters
$Chezy, Wx, Wy, CSHE, Teq, Rdecay, Lat.
```

Chezy	Wx	Wy	CSHE	Teq	Rdecay	Lat

```
$7.Output Profiles
$nprofiles     nprofiles
$ipwest, ipeast, jpsouth, jpnorth
$U-velocity, V-velocity, W-velocity
$nconstits
$nconst1, nconst2, nconst3, nconst4, etc
```

ipwest	ipeast	jpsouth	jpnorth
U-vel	V-vel	w-vel	
nconstits			
nconst1	nconst2	nconst3	nconst4

```
$8.Output Surfaces
$nsurfaces, nconstits      nsurfaces   nconstits
$U-velocity, V-velocity    U-vel       V-vel
$nconst1, nconst2, nconst3, nconst4, etc
```

nconst1	nconst2	nconst3	nconst4

```
$9.Output Time Series
$ntimser       ntimser
$nconst, iconst, jconst, kconst
```

nconst	iconst	jconst	kconst

```
$10.Simulation Time Conditions
$dtm, tmend            dtm       tmend
$tmeout,tmeserout      timeout   timeserout
$11.Internal Boundary Locations
$nintbnd       nintbnd
$ibwest, ibeast, jbsouth, jbnorth, ktop, kbottom
```

ibwest	ibeast	jbsouth	jbnorth	ktop	kbottom

```
$12. Constituent Averages
$nconarv       nconarv
$nconstarvs
```

nconst1	nconst2	nconst3	nconst4

```
$13 Groundwater Inflow
$ngrndwtr
$qgrndwtr  kgrndU  kgrndL
```

qgrndwtr	kgrndU	kgrndL

```
$temp,saln,const,bod_d,on_d,nh3,op_d,po4,phyt,bod_p,on_p,op_p
```

temp	saln	const	cbod	on	nh3	do(d)	no3	op	po4	phyt	bod_p	on_p	op_p

Table 2-2. Identification number, Nwqm, for each water quality model included in INTROGLLVHT and the constituent values that need to be specified for inflows, at elevation boundaries*.

		T	S	C	CBOD_D	DON_D	NH4_D	DO(D)	NO3_D	OP_D	PO4_D	PHYT	CBOD_P	ON_P	OP_P
WQM	nwqm/ncon	1	2	3	4	5	6	7	8	9	10	11	12	13	14
TSC	1	d	d	d											
DOD	2	d	d	d	d	d	d	d	d						
Eutro_P	3	d	d	d	d	d	d	d	d	d	d	d	d	d	d
Sed	4	d	d	d											

*The "d" indicates required data values. Those blanked out are not specified.

C, constituent; CBOD_D, dissolved carbonaceous biochemical oxygen demand; CBOD_P, particulate carbonaceous biochemical oxygen demand; DOD, dissolved oxygen deficit; DO(D),dissolved oxygen ; Eutro_P, Eutro5 particulate biochemical oxygen demand model (WQDPM) ; NH4_D,ammonium ; NO3_D,nitrate; ON_D,dissolved organic nitrogen ; ON_P ;particulate organic nitrogen OP_P, particulate organic phosphorus PO4_D,dissolved phospahte; PHYT,phytoplankton; S,salinity; Sed, sediment scour and deposition model; T,temperature ; TSC, temperature salinity arbitrary constituent model; Sed, sediment scour and deposition model, WQM, water quality model.

Table 2-3. Constituent numbers for specifying net algal growth rate, surface elevation, and velocity components for the time series output.

Constituent number (nconstit.)	Variable
15	Net algal growth rate
20	Elevation (specify k = 2)
21	U-velocity component
22	V-velocity component

Table 2-4. Example input file for Est_TSC_01 project.

```
Est_TSC_01
 $1.nwqm                 1
 $2.Inflow Conditions
 $ninflows               2
 $qinflow,iinflow,jinflow,kinflow
 $intake,inintake,jintake,kintake
 $temp,saln,const,cbod,on,nh3,do(d),no3,op,po4,phyt,bod_p,on_p,op_p
      5.00      2      4      2
      0        0      0      0
      0.00      0.00     0.00
      1.00     10      3      2
      1        8      2      2
      0.00      0.00    100.00
 $.4 Outflow Conditions
 $noutflows              1
 $qoutflow,ioutflow,joutflow,koutflow
      1.00      8      2      2
 $4.Elevation Boundary Conditions
 $nelevation   kts       1              2
 $iewest,ieeast,jesouth,jenorth
 zmean,zamp,tmelag,tideper
 k,temp,saln,const,cbod,on,nh3,do(d),no3,op,po4,phyt,bod_p,on_p,op_p for
 k=2,km-2
      20     20      3      5
      0.40      0.40      0.00      12.45
      2      0.00     10.00      0.00
      3      0.00     10.00      0.00
      4      0.00     10.00      0.00
      5      0.00     12.00      0.00
      6      0.00     15.00      0.00
      7      0.00     18.00      0.00
      8      0.00     20.00      0.00
      9      0.00     20.00      0.00
     10      0.00     20.00      0.00
     11      0.00     20.00      0.00
 $5.Initialize Water Quality Profiles
 $ninitial               1
 k,temp,saln,const,cbod,on,nh3,do(d),no3,op,po4,phyt,,bod_p,on_p,op_p for
 k=2,km-2
      2      0.00      0.00      0.00
      3      0.00      0.00      0.00
      4      0.00      0.00      0.00
      5      0.00      0.00      0.00
      6      0.00      0.00      0.00
      7      0.00      0.00      0.00
      8      0.00      0.00      0.00
      9      0.00      0.00      0.00
     10      0.00      0.00      0.00
     11      0.00      0.00      0.00
```

Table 2-4 (continued)

```
$6.External Parameters
 Chezy, Wx,Wy,CSHE, TEQ, Rdecay, Lat
    35.00     0.00     0.00    25.00      0.00      0.00     30.30
 $7.Output Profiles
 $nprofiles          1
 $ipwest,ipest,jpsouth,jpnorth
 $u-vel, v-vel, w-vel
 $nconstituents
 $I-const(1),I-const(2), I-const(3), etc
     2    20     4     4
     1     0     0
     2
     2     3
 $8.Output Surfaces
 $nsurfaces  $nconstituents            2            2
 $U-vel  V-vel   1.000000        1.000000
 $I-const(1), I-const(2), I-const(3), etc.
     2     3
 $9.Output Time Series
$ntimser        3
 $nconst, iconst, jconst,kconst
     3     8     2     2
     3    10     4     2
     2     2     4     5
 $10.Simulation time conditions
 $dtm   tmend   120.0000        480.0000
 $tmeout  tmeserout   1.556000        1.000000
 $11.Internal Boundary Locations
 $nintbnd          0
 $ibwest,ibeast,jbsouth,jbnorth,ktop,kbottom
 $12. Constituent Averages
 $nconarv          2
 $nconstarvs
     2    3
$13. Groundwater Inflow
 $ngrndwtr              0
```

3. EXECUTING THE MODEL, OUTPUT FILES, AND WATER QUALITY MODEL DEFAULT PARAMETERS

In this chapter, the steps required for executing the example application project set up in Chapter 1, Section 1.5 and Chapter 2, Section 2.3 are explained, followed by a description of what will appear on the screen while the model is running. Also covered are some diagnostics if the model does not run as set up. This is followed by an examination of the different output files, their content, and format.

For running the water quality models beyond the TSC model, default parameters for the models are required. The default parameter files for the DOD, WQDPM, and SED models are also presented and discussed in this chapter.

3.1 Executing the Model

With the EST_TSC_01_Con.dat, Estuary_BATH.dat, EST_TSC_01_INP.dat, and the skeleton EST_TSC_01_WQM.dat files available in the INTROGLLVHT Folder, the GLLVHT model is just about ready to run. Two steps are required before it can be run, however:

First, place the default values for the water quality model being run into the _WQM.dat file. For the TSC model all the default values have been specified as inputs and the EST_TSC_01_WQM.dat file should be left blank as it is. See Section 3.3 for the default files of the DOD, WQDPM, and SED water quality models.

Second, the file names for running the project should be transferred from the EST_TSC_01_Con.dat file to the ABControlfiles.dat file. Only the file names should be transferred, not the project notes made in the former.

The project can now be run by double-clicking on the GENERALMODEL.exe icon.

3.1.1 Screen Display During Execution

The first display on the screen is a list of the names of the bathymetry, input, and water quality model files entering the simulations. The simulation will pause so that these can be checked to ensure that the correct project is being run. This list is identical to the file names in ABControlfiles.dat. If they are the correct project files, type 1; if not, type 0 and the simulation will terminate.

During execution, the screen will display a number of parameters that aid in monitoring the simulation as it progresses. Shown on each time step are the following:

- the maximum time step dtm

- the actual time step of integration, dt, which should normally be 0.75 times dtm

- the cumulative simulation time, in hours

- the K level of the top layer. This should always be K = 2. If not, then there is a water imbalance as might occur in a reservoir case when the sum of the inflows do not equal the sum of the outflows. This is not usually a problem for tidal cases.

- the sum of the inflows when a reservoir case is being run

The main indicator of the status of the simulation is the actual time step, dt. One indication that the actual time step is too large is that it will degrade fairly rapidly to a very low value that is less than 0.1 of the original dtm. The actual time step may also vary between higher and lower values when tidal projects are run; however, if the two values keep decreasing, the dtm is too high. It is recommended that short simulations be run by setting tmend to a few hours to find a smaller time step.

3.1.2   Possible Execution Errors

The model may not run the first time it is executed, particularly if a new project is created by modifying existing project files. There are a number of common errors that can occur that will stop the execution of the model or cause it to show errors when it attempts to run. Possible execution errors include the following:

- The name of the files listed in the ABControlfiles.dat file may not correspond to available files. Compare file names carefully, looking for too many spaces or different spellings of parts of the names.

- There are extra lines of data or not enough lines of data in the _INP.dat file. Check the _INP.dat file carefully against the template in Table 2-1.

- The wrong water quality model identifier, nwqm, has been specified. The model is therefore looking for an incorrect number of water quality parameters.

- There are extra lines of data or not enough lines of data in the _WQM.dat file. Check the project file against the default water quality model being used.

- The KM may not be large enough. Reset KM to a larger value in the bathymetry file, and remember to add sufficient lines in the tidal boundary and initialization files so that they go from K = 2 to KM-2.

3.2 Description of Output Files

The output files generated by the EST_TSC_01 project simulation are presented here to provide a description of output file format and content. Only portions of each file are shown. The complete printout results can be examined from the output files that show up in the INTROGLLVHT Model folder at the end of the simulation. They are also stored in the Example Application folder in a subfolder named EST_TSC-01. Interpretation of the results describing the estuarine dynamics and constituent distributions can be found in Chapter 4.

The two tabular output files are the spatial output file, _SPO.dat, and the time series output file _TSO.dat. These are presented in a format that can be used to directly interpret the model results.

The format of the plot files are designed to be used directly in TECPLOT© software. For use in other plotting software such as MATLAB©, the files can be transferred to a spreadsheet for editing. Alternatively, a separate small program can be written that reads the files and edits them to the format required by other software.

3.2.1   Spatial Output File _SPO.dat

The spatial output file contains the profile slice results and the surface results for the conditions specified in Chapter 2, Section 2.1.7 and Section 2.1.8. The spatial output file will be output at a time interval of tmeout hours. For tidal cases, it will output results at the time interval of tmeout over the last two tidal cycles. This allows the spin-up of a tidal application to stationary state before the results are printed out.

For the estuary project set up in Chapter 2, the name of its spatial output file (SPO) file is EST_TSO_01_SPO.dat. This output file can become very lengthy if too many constituents are selected for display as profile slices or as surface results at too frequent a time interval.

A portion of the SPO file for the example project is shown in Table 3-1. The first part of the table is the K-levels, which gives the K0(I,J) values at each horizontal grid cell over the computational grid extending from I = 2 to IM-1 from left to right (west to east) and J = 2 to J = JM-1 from bottom to top (south to north). The zero values indicate land and the remaining values indicate water. This table is used to check that specified inflows and outflows are within the range of K = 2 to K(I,J) at their particular location. The K-level table enters the SPO file before the first time step is iterated. The computation can be stopped and the K0(I,J) values examined before a long simulation is performed.

The names of the project bathymetry, input, and water quality files used to generate the output are listed. They are given so that there is a record of the files that produced the output.

The simulation time of the profile and the surface output results that follow it is printed. The profile slice results along J = 4 are then shown. First, the water surface elevation along the slice is shown. Next, the U-velocity profiles at each I location along J = 4 extending from K = 2 to KM-2 are shown. This is followed by the profiles of salinity and the constituent (i.e., tracer dye) concentration. The profile tables for the velocity component and constituents can be copied to a spreadsheet for plotting individual profiles at specific I locations.

The surface and/or bottom output is for the velocity components and constituents specified in the _INP.dat file. They are also shown in Table 3-1. Examination of the full project example application SPO shows that the previous set of output is repeated for the next output time interval.

If a number of conarv constituents are specified, the last part of the _SPO.dat file will contain the velocity components specified for the profiles or surfaces, but only the constituents specified for the conarv. The velocity components and the constituent concentrations are averaged over the last two tidal cycles of the simulations.

3.2.2   Time Series Output File _TSO.dat

The time series output file contains the results for the conditions specified in Chapter 2, Section 2.1.9 and gives the concentrations for selected constituents at selected locations at each time interval of tmeserout hours. The time series file is named EST_TSC_01_TSO.dat, and a portion of it is shown in Table 3-2. The constituent values in the table are in exponential notation, which allows one to examine values that might be too large or small to show up in the _SPO.dat files.

The time series of constituent concentrations at a number of locations throughout a water body are used to determine if steady-state conditions have been reached for steady inflow outflow cases, or if time-varying stationary-state cases have been reached for time-varying tidal cases. The time series output file can be placed into a spreadsheet for time series plots of model results.

3.2.3   Plot Output Files

The plotter files contain the velocity vectors and all constituent concentrations of the particular water quality model being used at the grid locations for the specific longitudinal-vertical profiles that are printed out, and for the surface and/or bottom results that are printed out.

A limited portion of the profile plot file named EST_TSC_01_PLT_PO1.dat is shown in Table 3-3. In the name, PLT identifies the file as a plot file, and PO1 identifies it as a profile plot file for the first specified profile slice. A second profile slice would be named _P02.dat, and so on.

The first line in Table 3-3 gives the names of the variables in the columns. They are as follows: the x coordinate distance in meters from west to east, a z elevation measured from a zero reference level, the u and w velocity components for that location, and all of the water quality constituents in the water quality model specified. In the project being examined, it is the TSC water quality model and the constituents are as indicated in the heading: temperature, salinity, and the constituent concentration.

The second line gives the time of the plot data and the number of data points horizontally and vertically. A limited portion of the x–z plane follows the second line. A complete profile plot file is found in the EST_TSC_01_PLT_PO1.dat project file.

A profile specified from a value of jesouth to jenorth along a specific I value that is in the y–z plane would include the v and w velocity components and all of the constituents of the specified water quality model.

The surface or bottom plot files designated as EST_TSC_01_PLT_SUR.dat and EST_TSC_01_BOT.dat for the example application would contain the coordinates of the x–y plane, the u and v velocity components, and all of the constituents of the specified water quality model.

If a number of conarv constituents are specified, the last part of the plotter files will contain their appropriate coordinates and velocity components, but only the constituents specified for the conarv. This last file can be plotted to give the tidally averaged velocity vectors and constituent contours.

3.2.4   The _BATH_PLT.dat File

The Estuary_BATH_PLT.dat file contains the x and y coordinates at each I, J grid location and the depth at the center of the cell at that location. This file is updated each time a project run is made, but will only change if there has been a change in the bathymetry data file itself. The _BATH_PLT.dat file is used to make plots of the water body with contoured depths. The Estuary bathymetry plot is shown in Figure 3-1. It gives the depths to the nearest meter, and shows the remnant river channel, a reasonable representation of the estuary shape, and the island in the northeastern arm.

3.2.5   Plot Files for Animations

Plot files for animations may require output at small tmeout steps. It is helpful to produce these files without lengthy _SPO.dat files. This can be accomplished by specifying the number of profile slice files and giving only their orientation (i.e., iewest, ieeast, jesouth, jenorth), and not specifying that any velocity components or constituents be printed out.

The same can be done to get surface and or bottom plot files without generating lengthy _SPO.dat files. Specify a surface or surface and bottom and do not specify any velocity components or constituents. Also, do not specify any time series data output because the _TSO.dat file could get lengthy.

What shows up in the _SPO.dat file is the K0(I,J) table; the name of the bathymetry, input, and water quality model files used to generate the output; and a listing of the times of the output.

To avoid overwriting previous output, a new _INP.dat file should be set up with a new project name and the aforementioned specifications. The new _INP.dat file can be produced by copying the one for which the plotter animations are desired, changing the project name (which will change file names), updating the skeleton water quality file, and making the remaining changes in specifying the profile slices and the surface and or bottom output.

3.3 Water Quality Model Default Parameter Files

The water quality model files provide the rate coefficients and parameters required to run the model. Each model has a different set of rate coefficients and a different default parameter file format.

3.3.1   The Temperature, Salinity, First-Order Decay Consituent Model

The parameters used in the TSC model are the coefficient of surface heat exchange, the equilibrium temperature of surface heat exchange, and the first-order decay rate for the arbitrary constituent. All of these parameters are specified in the _INP.dat file because the TSC constituents are used in all water quality models. The default file for the TSC model in the INTROGLLVHT Model folder is named tsc_wqm.dat. This default file is blank because all of the TSC rate parameters have been specified.

Estimates of the coefficient of surface heat exchange and the equilibrium temperature are presented in Chapter 11. First-order decay rates for various substances are presented in Chapter 10.

3.3.2   The Dissolved Oxygen Depression Model

The default values of the parameters required to run the DOD model are found in the DOD WQM.dat file of the INTROGLLVHT Model folder. These default values should be copied into the project _WQM.dat file and modified for the project application.

The contents of the DOD WQM.dat default file are shown in Table 3-4. The rates are specified at 20° C and temperature effects are accounted for in the model. The table first identifies the water quality model number nwqm and should not be changed. The default values of the rate parameters and their definitions are given in the table. Further, definition of the rates are summarized in Chapter 12, where the derivations of the DOD relationships are given.

3.3.3   The Eutrophication Model

The default parameters for the WQDPM model are found in the file named WQDPM_WQM.dat of the INTROGLLVHT Model folder. The rates and constants required by the WQDPM model can be found in Table 13-3 of Chapter 13, where the WQDPM model is presented in detail. The major WQDPM parameters made available in INTROGLLVHT for adjustment are given in Table 3-5. The table lists the constituent equation being examined and its key parameter. One has the option to use the particular constituent equation or not. For example, to run a simple CBOD and dissolved oxygen balance, only the CBOD and DO equations are turned on and the rest are turned off. The key parameter for a dissolved constituent is its decay rate, and the key parameter for a particulate constituent is its settling rate. For each parameter, a default value is given, as well as a typical range of the parameter

There are many parameters related to the phytoplankton equation. First are the key rates of maximum growth rate, death rate and respiration rate, and the settling velocity. Next is the saturation light intensity for a particular alga species. Algae have the potential to produce the most oxygen at their saturation light intensity. After this is the assimilative efficiency of the zooplankton. An assimilative efficiency of close to unity means that the zooplankton will retain a particular constituent from the algae rather than pass it on to the particulate form of that constituent. An excretion fraction of carbon directly from the algae to dissolved CBOD is specified. Both the assimilative efficiency and the excretion fraction are included in Table 3-5, rather than given as fixed default values, because of their wide range of uncertainty. The default values of the zooplankton grazing rates given in Table 13-3 can be changed using the gzoo term. Setting gzoo to zero turns off zooplankton grazing.

There are two ways to evaluate sediment oxygen demand (SOD) and the release of nutrients from the sediments in WQDPM. The first method is to use Equation 13.6.4 in Table 13-2, similar to that developed by Pamatmat (1971) in which the SOD is a weighted sum of the rate of CBOD_P, ON_P, and phytoplankton carbon reaching the sediment. The rate at which each reaches the sediment is the product of the settling velocity and the concentration near the bottom. Each is weighted by the indicated fractions. The second method is to specify a measured SODm that is constant throughout the water body. In either case, the rate at which ammonium is released from the sediment can be taken proportional to the SOD using the spnh3 term (Pamatmat 1971). Alternatively, the spnh3 can be set to zero, and the measured SedNH3m specified. The rate of release of PO4 from the sediment also needs to be specified.

Last are the environmental parameters. The clear sky solar radiation and the cloud cover are used to compute the incident solar radiation to the surface of the water body. Clear sky solar radiation is given as a function of latitude and month in Table 11-1. The incident solar radiation is attenuated downward through the water column by a light extinction coefficient and the algal density as shown by Equation 13.4.3 in Table 13-2. The last environmental parameter is the reaeration wind speed. The Mackay (1980) surface reaeration formula is used in WQDPM. Further definition of the rates and parameters are given in Chapter 13, where the WQDPM relationships are presented.

3.3.4   *The Sediment Scour Deposition Model*

The default values of the parameters required to run the SED model are found in the Sed_WQM.dat file of the INTROGLLVHT Model folder. These default values should be copied into the project _WQM.dat file and modified for the project application.

The default parameters for the SED model are given in Table 3-6. They are the particle diameter (diasedmm), the specific gravity (spgr), the Shields parameter (Shldp), and the bottom minimum sediment concentration at which resuspension starts (cbotmin). The relationships between diasedmm, spgr, and cbotmin, as well as the range of Shldp, are discussed in Chapter 14.

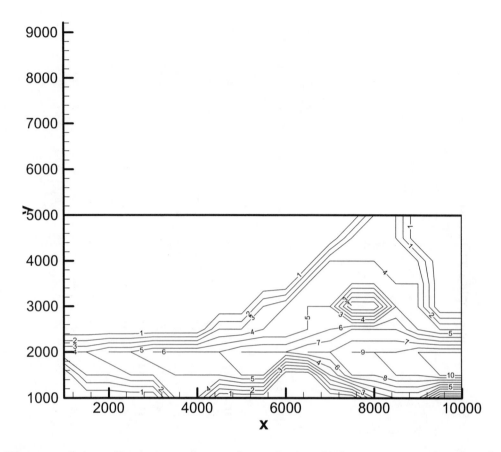

Figure   3-1.   Contour   map   of   estuary   bathymetry   produced   from Estuary_BATH_PLT.dat file. Axis distances and depths are given in meters.

Table 3-1. Portion of spatial output file for the EST_TSC_01 example application project.

```
K levels
 0 0 0 0 0 0 0 0 0 0 0
 0 0 0 0 0 0 0 0 0 0 0
 0 0 0 0 0 0 0 0 0 0 0
 0 0 0 0 0 0 0 5 6 8 8 11 11
 0 5 5 6 7 7 6 6 6 8 9 11 11
 5 4 4 5 6 6 7 8 9 10 10 11 11
 5 4 4 5 6 5 6 5 0 10 10  0  0
 0 0 0 0 0 0 0 0 0  6  6     0

Estuary_BATH.dat
EST_TSC_01_INP.dat
EST_TSC_01_WQM.dat

simulation time, hrs     456.6750

West-East water surface elevations, cm at j= 4
 59.16 58.99 58.85 58.73 58.64 58.56 58.51 58.47 58.46 58.45 58.46 58.46 58.45 58.46 58.45 58.46 58.45 58.41 58.35
 58.43 58.43

West-East U-velocity profiles, cm/sec at j= 4
     19.   20.    3.    4.    5.    6.    7.    8.    9.   10.   11.   12.   13.   14.   15.   16.   17.   18.
    3.50  3.77  4.02  4.04  4.08  2.83  2.59  2.31  1.62  2.45  1.52  0.16  0.03 -0.12 -0.17  1.02  5.76
    7.02  6.89  1.54  1.96  1.61  2.87  3.42  2.53  2.86  4.68  3.63  3.49  2.63  2.47  2.46  4.29  8.11
   -0.80 -0.13  0.15  1.12  2.77  3.96  5.84  4.14  4.42  6.48  6.28  6.97  5.72  5.48  6.19  7.00 10.40
    6.11  6.00  0.16  1.51  1.95  3.86  5.72  5.72  6.58  8.69  8.92 10.52 10.39  9.71  8.81  9.88 17.24
    6.00  0.60 -1.63 -3.72 -0.50 -0.50  0.61  3.34  5.51  9.90 10.31 11.05 10.39  9.05  8.15  9.97 14.14
    6.00  5.91  0.00  0.00 -1.84 -2.70 -2.23 -0.34  3.44  7.39  8.41  7.84  7.38  5.86  5.45  6.91 10.84
   -1.27 -2.48  0.00  0.00  0.00  0.00  0.00  0.00 -0.70  1.48  3.44  4.51  4.80  3.24  2.76  3.68  1.38
    8.09  8.00  0.00  0.00  0.00  0.00  0.00  0.00  0.00  0.00  0.00  1.18  1.52  1.28  0.80  1.62  0.40
    7.14  7.06  0.00  0.00  0.00  0.00  0.00  0.00  0.00  0.00  0.00  0.00  0.00 -0.99 -0.72 -1.18 -0.49
    4.76  4.71  0.00  0.00  0.00  0.00  0.00  0.00  0.00  0.00  0.00  0.00  0.00  0.00  0.00  0.00 -1.13
    0.98  0.96  0.00  0.00  0.00  0.00  0.00  0.00  0.00  0.00  0.00  0.00  0.00  0.00  0.00  0.00  0.00
    0.76  0.75  0.00  0.00  0.00  0.00  0.00  0.00  0.00  0.00  0.00  0.00  0.00  0.00  0.00  0.00  0.00
    0.63  0.63  0.00  0.00  0.00  0.00  0.00  0.00  0.00  0.00  0.00  0.00  0.00  0.00  0.00  0.00  0.00
    0.54  0.54  0.00  0.00  0.00  0.00  0.00  0.00  0.00  0.00  0.00  0.00  0.00  0.00  0.00  0.00  0.00
    0.00  0.00  0.00
```

Table 3-1 (continued)

West-East Constituent Profiles for saln at j= 7.[4]

19.	2.	3.	4.	5.	6.	7.	8.[4]	9.	10.	11.	12.	13.	14.	15.	16.	17.	18.
10.00	7.37	8.28	9.07	9.77	10.24	10.60	10.88	11.01	11.02	11.00	10.88	10.76	10.64	10.49	10.36	10.27	10.24
10.00	10.73	10.87	10.91	10.93	10.96	11.11	11.16	11.19	11.19	11.21	11.09	11.07	10.91	10.75	10.66	10.71	10.76
10.00	12.09	11.99	11.86	11.82	11.83	11.82	11.77	11.64	11.58	11.61	11.53	11.46	11.29	11.23	11.25	11.26	11.42
12.00	12.66	12.81	13.02	12.98	12.87	12.84	12.83	12.84	12.72	12.51	12.34	12.28	12.28	12.30	12.38	12.48	12.21
15.00	0.00	0.00	13.88	14.14	14.42	14.33	14.32	14.40	14.36	14.31	14.07	13.92	13.92	13.97	14.08	14.12	14.19
18.00	0.00	0.00	0.00	0.00	15.24	15.56	15.99	16.29	16.29	16.17	16.14	16.15	16.33	16.47	16.66	16.96	17.44
20.00	0.00	0.00	0.00	0.00	0.00	0.00	0.00	0.00	17.25	17.52	18.00	18.04	18.28	18.53	18.69	18.85	19.18
20.00	0.00	0.00	0.00	0.00	0.00	0.00	0.00	0.00	0.00	0.00	0.00	19.04	19.20	19.38	19.42	19.55	19.77
20.00	0.00	0.00	0.00	0.00	0.00	0.00	0.00	0.00	0.00	0.00	0.00	0.00	0.00	19.66	19.77	19.85	19.95
20.00	0.00	0.00	0.00	0.00	0.00	0.00	0.00	0.00	0.00	0.00	0.00	0.00	0.00	0.00	0.00	0.00	19.98
0.00	0.00	0.00	0.00	0.00	0.00	0.00	0.00	0.00	0.00	0.00	0.00	0.00	0.00	0.00	0.00	0.00	0.00

West-East Constituent Profiles for constat j= 7.[4]

19.	2.	3.	4.	5.	6.	7.	8.[4]	9.	10.	11.	12.	13.	14.	15.	16.	17.	18.
0.00	0.52	0.60	0.67	0.77	0.90	1.12	1.44	1.43	1.99	2.03	0.72	0.32	0.17	0.10	0.06	0.03	0.02
0.00	0.77	0.81	0.87	0.96	1.07	1.23	1.42	1.42	1.59	1.88	1.03	0.62	0.35	0.19	0.12	0.10	0.07
0.00	0.82	0.85	0.92	0.99	1.08	1.18	1.31	1.38	1.51	1.71	1.24	0.92	0.61	0.43	0.31	0.22	0.16
0.00	0.86	0.88	0.89	0.94	1.00	1.10	1.20	1.27	1.37	1.53	1.34	1.13	0.88	0.71	0.61	0.47	0.29
0.00	0.00	0.00	0.89	0.90	0.89	0.95	1.01	1.06	1.12	1.18	1.10	0.97	0.78	0.65	0.55	0.45	0.35
0.00	0.00	0.00	0.00	0.00	0.81	0.77	0.71	0.66	0.68	0.70	0.66	0.58	0.46	0.38	0.31	0.24	0.16
0.00	0.00	0.00	0.00	0.00	0.00	0.00	0.00	0.00	0.46	0.40	0.30	0.26	0.20	0.15	0.11	0.09	0.04
0.00	0.00	0.00	0.00	0.00	0.00	0.00	0.00	0.00	0.00	0.00	0.00	0.11	0.08	0.06	0.05	0.03	0.01
0.00	0.00	0.00	0.00	0.00	0.00	0.00	0.00	0.00	0.00	0.00	0.00	0.00	0.00	0.03	0.02	0.01	0.00
0.00	0.00	0.00	0.00	0.00	0.00	0.00	0.00	0.00	0.00	0.00	0.00	0.00	0.00	0.00	0.00	0.00	0.00
0.00	0.00	0.00	0.00	0.00	0.00	0.00	0.00	0.00	0.00	0.00	0.00	0.00	0.00	0.00	0.00	0.00	0.00

Table 3-1 (continued)

Surface west-East U-surface velocities, cm/sec

```
0.00  0.00  0.00  0.00  0.00  0.00  0.00  0.00  0.00  0.00
0.00  0.00  0.00  0.00  0.00  0.00  0.00  0.00  0.00  0.00
0.00  0.00  0.00  0.00  0.00  0.00  0.00  0.00  0.00  0.00
0.00  0.00  0.00  0.00  0.00  0.00  0.00  0.00  0.00  0.00
0.00  0.00  0.00  0.00  0.00  0.00  0.00  0.00  0.00  0.00
1.55  1.53  0.00  0.00  0.00  0.76  0.33  0.93  1.26  -0.07  -0.17
7.02  3.50  3.77  4.04  4.08  -0.26  0.37  1.62  2.31  2.59  2.83  3.02
0.00  6.89  1.99  3.13  4.11  5.02  1.52  2.45  3.20  3.21  2.80  3.62  5.76
6.10  5.92  0.00  0.00  0.88  4.63
0.00  0.00
```

Surface South-North V-surface velocities, cm/sec

```
0.00  0.00  0.00  0.00  0.00  0.00  0.00  0.00  0.00  0.00
0.00  0.00  0.00  0.00  0.00  0.00  0.00  0.00  0.00  0.00
0.00  0.00  0.00  0.00  0.00  0.00  0.00  0.00  0.00  0.00
0.00  0.00  0.00  0.00  0.00  0.00  0.00  0.00  0.00  0.00
0.00  0.00  0.00  0.00  0.00  0.00  0.72  1.26  0.81  1.78  0.98  0.61
0.00  0.00  0.00  0.00  0.00  0.00  0.00  1.46  0.38  2.14  1.49  3.11
2.13  -2.12  -1.68  -0.45  -0.92  -0.35  -0.51  0.85  1.07  -0.20  -0.38  -0.49
0.18  0.00  0.19  -0.53  -0.23  -0.33  1.12  0.49  -0.26  -1.17  -0.73  -1.15
0.00  0.00  -1.58  0.00  0.00  0.00  0.00  -1.94  -0.15  2.59
```

(right-side values: -1.82, -3.82, -5.14, -3.39, -0.38, 0.62, -1.44, 0.04, -0.49, -0.46, -1.10, -2.41, -1.06, -2.39, -1.06)

Surface Constituent for saln

```
0.00   0.00   0.00   0.00   0.00   0.00   0.00   0.00   0.00   0.00
0.00   0.00   0.00   0.00   0.00   0.00   0.00   0.00   0.00   0.00
0.00   0.00   0.00   0.00   0.00   0.00   0.00   0.00   0.00   0.00
0.00   0.00   0.00   0.00   0.00   0.00   0.00   0.00   0.00   0.00
0.00   0.00   0.00   0.00   0.00   10.95  10.87  11.01  11.06  10.36  10.36
7.37   8.28   0.00   0.00   0.00   10.87  10.80  11.02  11.01  10.42  10.49
10.00  10.00  9.27   9.77   10.24  10.60  10.88  11.00  10.98  10.89  10.88  10.86  10.27
10.00  10.00  9.07   9.74   10.32  10.59  10.89  11.07  11.02  10.90  10.86  10.83  10.77
10.00  10.00  0.00   0.00   10.83  10.53  10.44  10.49  10.42
0.00   0.00
```

(right-side saln values: 10.68, 10.53, 10.39, 10.25, 10.24, 10.22, 10.42, 10.70, 10.57, 10.29, 10.32, 10.49, 10.71, 10.61, 10.64, 10.61, 10.79, 10.68, 10.76)

Table 3-1 (continued)

Surface Constituent for const

```
0.00  0.00  0.00  0.00  0.00  0.00  0.00  0.00  0.00  0.00  0.00  0.00  0.00  0.00  0.15  0.14  0.00
0.00  0.00  0.00  0.00  0.00  0.00  0.00  0.00  0.00  0.00  0.00  0.00  0.00  0.14  0.14  0.12  0.00
0.00  0.00  0.00  0.00  0.00  0.00  0.00  0.00  0.00  0.00  0.00  0.00  0.17  0.15  0.14  0.11  0.10
0.00  0.00  0.00  0.00  0.00  0.00  0.00  0.00  0.00  0.00  0.00  0.24  0.19  0.17  0.14  0.09  0.08
0.00  0.00  0.00  0.00  0.00  0.00  0.47  0.00  0.00  0.90  1.12  0.24  0.18  0.00  0.00  0.05  0.05
0.00  0.00  0.00  0.52  0.67  0.77  0.63  0.82  1.05  0.88  1.10  0.21  0.15  0.07  0.05  0.03  0.03
0.00  0.00  0.00  0.60  0.73  0.79  2.03  1.99  1.43  1.44  1.10  0.32  0.17  0.10  0.06  0.03  0.02
0.00  0.00  0.00  0.67  0.00  0.00  4.54  6.61  3.13  1.86  1.22  0.00  0.13  0.10  0.07  0.04  0.02
0.00  0.00  0.00  0.00  0.00  0.00  0.00  0.00  0.00  1.64  0.00  0.00  0.00  0.00  0.09  0.08  0.06
```

Bottom West-East U-bottom velocities, cm/sec

```
0.00  0.00  0.00  0.00  0.00  0.00  0.00  0.00  0.00  0.00  0.00  0.00  0.43  1.53  0.00
0.00  0.00  0.00  0.00  0.00  0.00  0.00  0.00  0.00  0.00  1.10  2.28  2.00  0.00  1.33
0.00  0.00  0.00  0.00  0.00  0.00  2.31  1.86  0.00  4.53  1.52  5.39  0.87  -0.72  1.44  3.95
-1.27  0.54  1.07  0.00  0.00  1.65  2.27  5.95  1.18  0.00  -0.99  0.44  0.00
3.69  3.69  0.54  -1.63  -1.10  5.62  6.90  3.44  0.00
-2.48  -0.10  -3.72  -0.09  4.80  3.63  1.48  0.00
-1.84  -0.54  4.64  -0.70  0.56  0.00
-2.23  0.77  1.32  -0.34  1.73  0.00
-2.70  -0.37  0.35
```

Bottom South-North V-bottom velocities, cm/sec

```
0.00  0.00  0.00  0.00  0.00  0.00  0.00  0.00  0.00  0.00  0.00  0.00  0.00  -1.05  0.00
0.00  0.00  0.00  0.00  0.00  0.00  0.00  0.00  0.00  0.00  0.00  0.00  -0.10  -0.49  0.00
0.57  0.47  0.00  0.00  0.00  0.00  1.11  0.00  0.00  0.00  -0.96  0.00  0.00  0.00  0.00  0.00
-1.80  -2.64  -2.55  -0.61  1.63  0.00  0.00  -1.96  0.00  -3.38  0.60  2.93  -0.28  0.82  0.75
-0.39  -0.39  -1.35  1.37  0.00  0.00  0.00  0.00  0.00  0.00  -1.31  0.33  4.67
0.00  0.00  0.00  0.00  0.00  0.00  -0.25  -4.18
```

45

Table 3-1 (continued)

Bottom Constituent for   saln

0.00	0.00	0.00	0.00	0.00				
0.00	0.00	0.00	0.00	0.00				
0.00	0.00	0.00	0.00	0.00				
0.00	0.00	0.00	0.00	0.00				
0.00	0.00	0.00	0.00	0.00				
12.00	12.00							
12.66	12.81	13.88						
20.00	20.00	11.76						
0.00	20.00	0.00						
0.00	0.00							

0.00	0.00	11.36	12.09	12.06	14.29	19.98	19.93	14.16
11.17	11.24	11.45	11.88	13.76	18.74	19.85	19.84	14.27
11.19	11.30	11.94	11.97	0.00	18.74	19.77	19.48	12.51
0.00	11.32	12.11	11.99	0.00	18.65	19.66	19.46	0.00
0.00	0.00	12.27	12.32	13.79	15.66	19.20	14.07	0.00
0.00	0.00	0.00	12.39	13.78	13.60	19.04	0.00	0.00
0.00	0.00	0.00	12.26	12.40	18.00	0.00	0.00	0.00
0.00	0.00	0.00	11.57	12.60	17.52	16.28	0.00	0.00
0.00	0.00	0.00	0.00	11.65	17.25	16.40	0.00	0.00
0.00	0.00	0.00	0.00	11.59	16.29	14.38	0.00	0.00
0.00	0.00	0.00	0.00	0.00	15.99	14.26	12.76	0.00
0.00	0.00	0.00	0.00	0.00	15.56	14.16	12.82	0.00
0.00	0.00	0.00	0.00	0.00	15.24	12.84	0.00	0.00
0.00	0.00	0.00	0.00	0.00	14.14	12.76	0.00	0.00

Bottom Constituent for   const

0.00	0.00	0.00	0.00	0.00				
0.00	0.00	0.00	0.00	0.00				
0.00	0.00	0.00	0.00	0.00				
0.00	0.00	0.00	0.00	0.00				
0.00	0.00	0.00	0.00	0.00				
0.86	0.88	0.90	0.81	0.77	0.71			
0.00	0.00	0.95	1.11	1.00	1.06			
0.00	0.85	0.96	0.00	1.17	1.28			
0.00	0.00							

0.00	0.00	0.20	0.23	0.21	0.33	0.00	0.00	0.30
0.20	0.21	0.24	0.25	0.28	0.10	0.01	0.01	0.38
0.20	0.22	0.34	0.33	0.00	0.11	0.02	0.04	0.38
0.00	0.23	0.38	0.36	0.00	0.12	0.03	0.05	0.00
0.00	0.00	0.41	0.43	0.48	0.40	0.08	0.61	0.00
0.00	0.00	0.00	0.47	0.53	0.64	0.11	0.00	0.00
0.00	0.00	0.00	0.00	0.64	1.02	0.30	0.00	0.00
0.00	0.00	0.00	0.00	0.77	1.31	0.40	0.72	0.00
0.00	0.00	0.00	0.00	0.00	1.25	0.46	0.67	0.00
0.00	0.00	0.00	0.00	0.00	1.32	0.66	1.15	0.00

Table 3-2. Portion of time series output file for the EST_TSC_01.

```
Estuary_BATH.dat
 EST_TSC_01_INP.dat
 EST_TSC_01_WQM.dat
Const.NM.           const        const        saln
I-Location           8            10           2
J-Location           2            4            4
K-Location           2            2            5
Time, hrs
   0.1025E+01    0.7029E-04   0.8666E-02   0.4961E-29
   0.2050E+01    0.1268E-02   0.1012E-01   0.3355E-20
   0.3075E+01    0.6802E-02   0.1306E-01   0.2076E-14
   0.4100E+01    0.1590E-01   0.3026E+00   0.2619E-11
   0.5125E+01    0.1942E-01   0.4482E+00   0.1131E-09
   0.6150E+01    0.1614E-01   0.3854E+00   0.1235E-08
   0.7175E+01    0.1302E-01   0.3153E+00   0.2606E-07
   0.8200E+01    0.1087E-01   0.2609E+00   0.1512E-05
   0.9225E+01    0.9920E-02   0.2250E+00   0.1407E-03
   0.1025E+02    0.9859E-02   0.2101E+00   0.6158E-02
   0.1127E+02    0.1028E-01   0.2194E+00   0.1942E+00
   0.1230E+02    0.9843E-02   0.2333E+00   0.7794E+00
   0.1332E+02    0.8830E-02   0.2211E+00   0.1734E+01
   0.1435E+02    0.8250E-02   0.2844E+00   0.3134E+01
   0.1537E+02    0.7806E-02   0.3378E+00   0.4416E+01
   0.1640E+02    0.7856E-02   0.3045E+00   0.5494E+01
   0.1742E+02    0.8587E-02   0.2658E+00   0.6454E+01
   0.1845E+02    0.9428E-02   0.2613E+00   0.7325E+01
   0.1947E+02    0.9760E-02   0.2905E+00   0.8120E+01
   0.2050E+02    0.9799E-02   0.3212E+00   0.8854E+01
   0.2152E+02    0.9829E-02   0.3589E+00   0.9547E+01
   0.2255E+02    0.9864E-02   0.4302E+00   0.1013E+02
   0.2357E+02    0.9976E-02   0.5535E+00   0.1051E+02
   0.2460E+02    0.1058E-01   0.6958E+00   0.1073E+02
   0.2562E+02    0.1173E-01   0.8278E+00   0.1085E+02
   0.2665E+02    0.1324E-01   0.7898E+00   0.1094E+02
```

Table 3-3. Portion of plot file for a profile slice for the EST_TSC_01 example application project

```
variables="x","z","u","w","temp ","saln ","const"
 zone T="        1    456.6750    Hr.",I=        19,J=
11,F=Point
```

x	z	u	w	temp	saln	const
1000.0000	10.0000	3.4984	0.0413	0.0000	7.3676	0.5231
1500.0000	10.0000	3.7733	0.0121	0.0000	8.2848	0.5961
2000.0000	10.0000	4.0163	-0.0128	0.0000	9.0711	0.6715
2500.0000	10.0000	4.0356	-0.0347	0.0000	9.7719	0.7718
3000.0000	10.0000	4.0758	-0.0376	0.0000	10.2368	0.8990
3500.0000	10.0000	2.8322	-0.0724	0.0000	10.6005	1.1153
4000.0000	10.0000	2.5896	-0.0413	0.0000	10.8753	1.4411
4500.0000	10.0000	2.3101	-0.0146	0.0000	11.0072	1.4293
5000.0000	10.0000	1.6201	-0.0797	0.0000	11.0244	1.9896
5500.0000	10.0000	2.4468	-0.0745	0.0000	11.0012	2.0289
6000.0000	10.0000	1.5217	-0.0666	0.0000	10.8808	0.7233
6500.0000	10.0000	0.1602	-0.0739	0.0000	10.7619	0.3172
7000.0000	10.0000	0.0312	-0.0765	0.0000	10.6354	0.1701
7500.0000	10.0000	-0.1219	-0.0574	0.0000	10.4878	0.0997
8000.0000	10.0000	-0.1693	-0.0301	0.0000	10.3570	0.0556
8500.0000	10.0000	1.0195	0.0015	0.0000	10.2672	0.0296
9000.0000	10.0000	5.7625	0.0052	0.0000	10.2437	0.0183
9500.0000	10.0000	7.0173	0.5384	0.0000	10.0000	0.0000
10000.0000	10.0000	6.8859	0.0260	0.0000	10.0000	0.0000
1000.0000	9.0000	-0.7954	0.0254	0.0000	10.7294	0.7663
1500.0000	9.0000	-0.1251	-0.0239	0.0000	10.8686	0.8079
2000.0000	9.0000	1.5370	0.0003	0.0000	10.9090	0.8680
2500.0000	9.0000	1.9586	-0.0236	0.0000	10.9273	0.9640
3000.0000	9.0000	1.6108	-0.0419	0.0000	10.9561	1.0662
3500.0000	9.0000	2.8725	-0.0467	0.0000	11.1096	1.2343
4000.0000	9.0000	3.4216	-0.0441	0.0000	11.1579	1.4213
4500.0000	9.0000	2.5305	-0.0467	0.0000	11.1907	1.4193

Table 3-4. Default values of the rate parameters for the dissolved oxygen deficit water quality model

Parameter	Value	
$nwqm	2	Model number
$kbod	0.15	Decay rate, per day
$knh	0.05	Ammonium decay rate, per day
$kon	0.10	Organic nitrogen decay rate, per day
$WAD	1.5	Reaeration wind speed, m/s

Table 3-5. Key Parameters in water quality dissolved particulate model water quality model.

Parameter	Default	Range	Description	Units
NH4				
NH4_yes	1	0 or 1		
K12	0.15	0.09-0.20	NH4 decay rate	day^{-1}
ON_D				
OND_yes	1	0 or 1		
K71	0.075	0.01-0.15	ON_D decay rate	day^{-1}
ON_P				
ONP_yes	1	0 or 1		
Vs7	0.08	0.05-0.20	ON_P settling rate	m/day
NO3				
NO3_yes	1	0 or 1		
K2d	0.09	0.09-0.16	Denitrification rate	day^{-1}
PO4				
PO4_yes	1	0 or 1		
BOD_D				
BODD_yes	1	0 or 1		
kd	0.15	0.02-0.20	BOD_D decay rate	day^{-1}
BOD_P				
BODP_yes	1	0 or 1		
Vs5	0.07	0.05-0.20	BOD_P settling rate	m/day
DO				
DO_yes	1	0 or 1		
OP_D				
OPD_yes	1	0 or 1		
K83	0.22	0.10-0.30	OP_D decay rate	day^{-1}
OP_P				
OPP_yes	1	0 or 1		
Vs8	0.08	0.05-0.20	OP_P settling rate	

Table 3-5 (continued)

Parameter	Default	Range	Term	Units
PHYT				
Phyt_yes	1	0 or 1		
K1c	2.0	1.5-3.5	Maximum growth rate	day^{-1}
K1d	0.3	0.2-0.6	Death rate	day^{-1}
K1r	0.07	0.05-0.2	Respiration rate	day^{-1}
Vs4	0.09	0.05-0.50	Settling velocity	m/day
Is	90.0	30-150	Sat. light intensity	W/m^2
As	0.70	0.5-0.9	Assim. effic. zoo.	
fe	0.10	0.1-0.8	Excret. Fraction phyto	
gzoo	1.0	0.0-2.5	Mult def graze rate	
Sediment				
S4	0.02	0.01-0.09	Fraction settled phyto to SOD	
SP5	0.07	0.01-0.09	Fraction settled BOD_P to SOD	
SP7	0.05	0.01-0.09	Fraction settled ON_P to SOD	
Spnh3	0.45	0.2-0.6	Fraction SOD released as NH3	
SODm	0.0	0.5-5.0	Measured SOD	g-O$_2$/ m^2/day
SedNH3m	0.0	0.3-3.0	Measured NH3 release from sediment	g-N/ m^2/day
SedPO4m	0.2	0.01-1.5	Measured PO4 release from sediment	g-P / m^2/day
Environ.				
Hsc	100.0	50-350	Clear sky solar rad. (See Table 11-1)	
Cloud Cover	6.0	1-10	Cloud cover	Tenths
Wad	2.5	0.5-5.0	Reaeration wind speed .	m/s

BOD_D, dissolved particulate carbonaceous biochemical oxygen demand; BOD_P, particulate carbonaceous biochemical oxygen demand; NH$_3$, ammonia; NH$_4$, ammonium; O$_2$, oxygen; ON_D,dissolved organic nitrogen ; ON_P,particulate organic nitrogen ; PO4, phosphate; SOD, sediment oxygen demand.

Table 3-6. Default parameters for the sediment scour and deposition model.

Parameter	Default Value	Definition and Units
$diasedmm	0.02	Sediment particle diameter in mm
$spgr	1.2	Particle specific gravity
$Shldp	0.08	Shields parameter
$cbotmin	1.0	Bottom minimum sediment concentration, mg/l

4. APPLICATION OF THE TEMPERATURE, SALINITY, FIRST-ORDER DECAY CONSTITUENT MODEL TO ESTUARIES AND COASTAL WATERS

The first application of INTROGLLVHT is for the estuarine project Est_TSC_01 for which the bathymetry was set up in Chapter 1, Section 1.5; the input file was set up in Chapter 2, Section 2.3; and which was then executed in Chapter 3, Section 3.1. The TSC model should generally be applied first to a project to make sure that the bathymetry is set up correctly, to ensure that the model will run for the specified input file conditions, and to examine the model spatial output file (SPO) and the time series output file (TSO) output results for circulation patterns to determine if they are reasonable.

Some of the TSC output for the example application project EST_TSC_01 is examined to illustrate what circulation looks like in a drowned river mouth estuary with freshwater inflow at its head and salinity stratification at its tidal mouth. A second estuary application project, EST_TSC_02, is set up to illustrate the data entering the input file so that residence times can be output through the length of the estuary. A third project, EST_TSC_03, is set up to illustrate the inclusion of groundwater inflow in the input file and its influence on salinity and velocity distributions.

An example open coastal project, Coastal_Brine_Discharge_01, is used to illustrate the setup of open coastal boundary conditions in the input file. This project is designed to examine the offshore distribution of a brine discharge from a distillation facility.

The input file and water quality file are presented for each project, along with the bathymetry file if it was not displayed previously. Only summary results of the model outputs are presented in this chapter. The full set of files for the example projects are saved in the Example Applications folder as individual subfolders. The latter may be referred to when describing additional detail about an example project. Additional examples that the user can try are listed with each application.

4.1 The Estuarine EST_TSC_01 Application

The complete set of input and output files for the example application project EST_TSC_01 are saved in the Example Applications folder as a subfolder named 4.1 EST_TSC_01 and can be viewed there.

4.1.1 *Approach to Stationary-State Conditions*

As shown in Table 2-4, the project simulation extends over 480 hours or 20 days to bring it to a stationary state where constituent concentrations, salinity, and velocities repeat themselves from tidal cycle to tidal cycle. The approach to stationary-state conditions can be determined from the time series output results, a portion of which is shown for this project in Table 3-2, and for which the full table is given in the project subfolder as the file named EST_TSC_01_TSO.dat.

One indicator included in the time series results used to measure the approach to stationary-state conditions is the salinity at the head of the estuary for location $I = 2$, $J = 4$, and $K = 5$. The project input table shows that the salinity is initialized from zero and its distribution builds up over time. This would take longest near the head of the estuary. The full time series table in the Example Application subfolder shows that the salinity does reach stationary state at the head of the estuary over the simulation time. The time series of salinity at the upestuary location is shown in Figure 4-1, which was prepared by placing the time series file in a spreadsheet. Figure 4-1 shows that salinity at the head of the estuary reaches stationary-state conditions in approximately 100 hr.

The facility discharge is connected to an intake within the water body and recirculation can build up. The second indicator of the approach to stationary-state conditions is to examine if the constituent dye concentrations off the intake and discharge repeat themselves over the last few tidal cycles. The locations of $I = 8$, $J = 2$, and $K = 2$ near the intake and $I = 10$, $J = 4$, and $K = 2$ near the discharge are chosen for examination of the dye concentration time series. The complete time series shows that the dye concentrations do reach stationary-state conditions over the simulation time.

Another important indicator of the approach to stationary-state is the velocity time series as output in the next project application EST_TSC_02. Figure 4-2 shows the U velocity component at the $I = 8$, $J = 2$, $K = 2$ location. It indicates that it takes up to 300 hr for the velocity to reach stationary-state conditions at this location.

4.1.2   Salinity Stratification and Circulation

The tidally averaged salinity stratification and velocity profiles are first examined for this project. The tidally averaged values are simply the values of the parameters averaged over the last two tidal cycles. The tidally averaged values for the chosen profile and surface display and chosen parameters are shown at the end of the spatial output table, EST_TSC_01_SPO.dat. The longitudinal-vertical profiles of salinity and velocity are extracted from the latter and are shown in Table 4-1 for a slice along $J = 4$.

The salinity and tidal boundary conditions apply at $I = 19$ and $I = 20$. The salinity profiles at the boundary are as specified for the tidal boundary condition in the project input file. The velocity values at the tidal boundary are fictitious, and the first real velocity profile generated by the inputs is at $I = 18$. Examination of the salinity profiles show that there is less salinity stratification up the estuary than at the mouth. The average salinity over the top layers at stations upestuary from the mouth are greater than at the mouth; for example, the average salinity over the top four layers at $I = 14$ is 10.78 ppt, whereas at the mouth it is 10.50 ppt. The less dense water in the surface layers at the mouth tends to push inward over the more dense water in the surface layers upestuary. In the bottom layers, the more dense water at the mouth tends to push inward under the less dense bottom waters upestuary.

This density-induced, or baroclinic, flow shows up in the velocity profiles given in Table 4-1, where there are inflows in the bottom and top layers. These flows, along the freshwater inflow at the head of the estuary, pass outward in the middle layers through which the average salinities are more uniform up and down the estuary. A tidally averaged three-layered flow is not uncommon in estuaries. It tends to be stronger where there is no freshwater inflow when the upestuary salinity profiles near the head of the estuary have little or no stratification and an average salinity nearly equal to that at the mouth. They tend to disappear as the freshwater inflow increases.

4.1.3 *The Tracer Dye Distributions and Surface Circulation*

The tidally averaged surface distribution of the tracer dye discharge, sometimes called a virtual tracer in simulations, is shown in Table 4-2. The surface distribution extends from $I = 2$ to $I = 20$ west to east and $J = 2$ to $J = 10$ from south to north. The pattern of the dye concentrations shows the distribution of the discharge plume. The increase in dye concentration across the facility was set at 100 ug/l.

The ratio of the increase in dye concentration across the facility to the dye concentrations on the surface indicates the amount of dilution the discharge undergoes. For the model cell receiving the discharge ($I = 10$, $J = 3$, $K = 2$), the dilution is 100/6.8 or approximately 15:1. The discharge dilutes very rapidly on the surface outward from the discharge. The tidally averaged dye concentration profiles along the $J = 4$ slice are also shown in Table 4-2. The profiles indicate that there is almost 10:1 additional dilution between the surface and bottom layers.

The surface west to east U-velocities are given in Table 4-2. It clearly shows the freshwater inflow moving downestuary to about $I = 8$ and mixing downward after entraining salinity. There is a complex surface flow around the discharge and intake area and around the island which extends from $I = 15$ to 16 at $J = 6$.

Both the surface and profile velocity fields show that the circulation within a water body can be quite complex. In an estuary, most of the transport of any water quality constituent is primarily determined by the tidally averaged flow. The distribution of water quality constituents can only be determined to the detail with which the flow field is known.

4.1.4 *Additional Suggested Study Examples*

A description of the results of the EST_TSC_01 example application project suggests a number of examples. These can be set up by making a copy of the EST_TSC_01_INP.dat input file and renaming it as indicated subsequently. The EST_TSC_01_CON.dat file should also be copied and renamed to keep track of the new project files needed to go into the master ABControl.dat file, and can be used as well to make notes about the example simulation.

The circulation within an estuary can vary depending on the freshwater inflow and tidal conditions. Pritchard (1952, 1955, 1956, 1967, 1969) showed that a relatively deep estuary can vary from a salt wedge estuary to a partially mixed estuary or a fully mixed estuary as these conditions vary seasonally. Many of the following suggested study examples are designed to illustrate different salinity and circulation patterns that can develop in an estuary.

Project EST_TSC_01a. Set up a project to try different freshwater inflow rates to see how the salinity and velocity profiles vary. Remember to change the name in the copy of the EST_TSC_01.INP.dat file that was made, and to change the input file name in the _CON.dat file. Include the time series of some U-velocity components at a few locations to determine how fast the velocity field becomes established. Remember to set the ntimser to the correct number. Vary the freshwater inflow rate and the mean tide and tidal amplitude levels to determine their influence on salinity structure and circulation. At what combination of conditions does a salt wedge estuary become established? Is there a combination of conditions for which it approaches a fully mixed estuary?

Project EST_TSC_01b. Set up a project to try different amounts of salinity stratification at the mouth of the estuary, particularly a salinity profile that is uniform from top to bottom. Run this case for different freshwater inflows to determine what happens to the circulation and dilution of the discharge. Is the boundary salinity profile important in determining the type of circulation pattern that becomes established?

Project EST_TSC_01c. Set up a project to determine the effects of recirculation on the virtual dye distributions. Disconnect the facility discharge from the intake. The outflow (withdrawal) from the estuary should also be turned off.

Project EST_TSC_01d. Set up a project with zero tidal amplitude at the boundary to determine how long it takes for the salinity profiles and velocity profiles to develop for these steady-state conditions. Determine how different the salinity distribution, flow field, and constituent distribution is for running the model for steady-state conditions versus running it for time varying tidal conditions.

Project EST_TSC_01e. Set up a project with a surface wind speed component beginning from the original EST_TSC_01 project. Compare the velocity fields and constituent distribution with and without wind on the surface.

4.2 Determination of Residence and Flushing Times

The flushing time or residence time at different locations throughout a water body is a very useful parameter for describing a water body. It indicates locations within the water body where circulation is strong and weak. The flushing or residence time is an estimate of how long it takes an initial concentration of dye within the water body to reduce to one half its concentration. The residence time can be compared with the half-life of a decaying material or with a half-life computed for a water quality kinetic reaction rate. For the latter, if the residence time is long in comparison to the reaction rate half-life, then most of the reaction would be completed at that location. The theoretical basis for computing residence time is presented in Chapter 10. The half-life computation is also included in the chapter.

The flushing or residence time can be computed for each cell in the model using the arbitrary constituent in the TSC water quality model. The computation is performed by having no constituent dye concentrations in any of the inflows, setting the Rdecay parameter to zero, and initializing the dye concentration to the same value, usually 100 ug/l, throughout the water body. The flushing or residence time is computed from the concentration of dye at each location from the concentrations at the end of a simulation.

4.2.1 The EST_TSC_02 Input and Control Files

The project EST_TSC_01 is set up to compute flushing or residence times as project EST_TSC_02. The input file for this project is shown in Table 4-3. The water quality model is nwqm = 1, as it was previously. The inflows are left the same and recirculation is included for the facility discharge, however, there is no dye in the facility discharge because the dye is used for estimating the residence time throughout the water body. The elevation boundary conditions are the same. The water quality profiles are initialized, including placing 100 ug/l of dye uniformly throughout the water column for all of the water body cells. External parameters remain the same.

One output profile is indicated for the same slice as previously, and with U and W velocity components displayed. No constituents are displayed with the profile; however, they could be. A surface is displayed, with no velocity components or constituents called for. There are three time series, including the time series of the constituent dye near the intake, the U velocity component (constituent 21) at $I = 10$, $J = 4$, and $K = 2$, and salinity at the head of the estuary. The maximum time step is 120 sec, and there is only one output at the end of the simulation (Tmend = tmeout), which is 480 hr. The time series are output every hour as before.

Two constituent averages are called for. They are salinity, which is constituent 2, and the dye as converted to residence time, which is 3. It is necessary to specify the latter in the constarv to get the residence time to print out. The input file is named Est_TSC_02_INP.dat.

The same bathymetry file, Estuary_BATH.dat, is used. If the project input file is generated using the INTROGLLVHT Input File.exe routine, then a new skeleton water quality model file is generated named Est_TSC_02_WQM.dat, which can be left blank to run the TSC model. If the project Est_TSC_02 input file is generated by copying and modifying the Est_TSC_01 project input file, then the Est_TSC_01_WQM.dat file could be used. The ABControls.dat file for the project should be as shown in Table 4-4. A copy to the ABControls.dat file should be saved as Est_TSC_02_CON.dat and stored with the remaining project files in the Example Applications subfolder.

4.2.2   Residence Time Results

The profile and surface distribution of residence time are shown in Table 4-5. The residence time is lowest near the mouth of the estuary and increases the further the distance from the mouth. They tend to be lower at the bottom of the water column than near the surface, indicating faster exchange of the bottom water out of the estuary.

The maximum values of about 5 to 6 days are found in the vicinity of the facility discharge and up toward the head of the estuary. These higher values are in the upper 2 m of the water column. The facility discharge and recirculation appear to be limiting flushing in that region. Another area of high residence time is to the north around the island.

The spatial output (SPO) file contains the average U velocity component for comparison to the previous results, and the vertical W velocity component. The latter can be examined to see the upwelling of water from the bottom layers to the middepth outflowing layers, and the downwelling of water from the surface layers to the middepth outflowing layers.

The time series output file includes the results of the dye concentrations over time at the location I = 8, J = 2, and K = 2. It shows the dye concentration decreasing with time at this location due to flushing. It also includes the time series of the horizontal velocity component at one location. The time series file is placed in a spreadsheet; Figure 4-3 shows the results of plotting the file. The figure shows the decrease in dye concentration at the chosen location, as well as an exponential decay fit. Computation of the half-life from the indicated equation gives a value of 5.5 days. This compares favorably with the results given in the surface residence time shown in Table 4-5.

4.2.3   Additional Suggested Study Examples

Following are some additional suggested study examples.

EST_TSC_02a. Turn off the facility discharge and intake (setting the facility inflow and outflow to zero should accomplish this) to determine if the recirculation between the intake and discharge, as well as the discharge flow itself, affects residence time in the estuary in its vicinity.

EST_TSC_02b. Set up a project to study the effects of different freshwater inflow rates at the head of the estuary on the residence or flushing time.

EST_TSC_02c. Set the tidal amplitude to zero to run a steady-state solution. Compare the residence times computed by this technique with those from the tidally varying project.

4.3 Estuary with Groundwater Inflow

The groundwater flow into a water body can be an important contribution to the body's water budget and water quality. It can have an effect on salinity and circulation as a relatively fresh water inflow into an estuary. Goetchius (2000) shows that groundwater is a major source of trace organics into estuaries, and it is necessary to include the groundwater inflow in the input data for the GLLVHT model.

An example application of the significance of groundwater into an estuary is provided in the project Est_TSC_03. The portion of the Est_TSC_01 project input file extended to include a groundwater inflow is given in Table 4-6. Item 13 in Table 4-6 shows that a groundwater inflow is specified, that the total groundwater inflow is 20 m^3/s, and that it enters the estuary into the bottom cells that have K0(I,J) at and between kgrndU = 3 and kgrndL = 4. The bathymetry K0(I,J) levels given in Est_TSC_01_SPO.dat show that the cells with bottom K0(I,J) in this range are located mostly north of the island, and upestuary near the river inflow. The groundwater inflow has a dye tracer concentration of 1000 ug/l to measure its dilution in the estuary. The output profiles are for the longitudinal-vertical slice along the centerline of the estuary, and a slice across the estuary just west of the island.

4.3.1 Groundwater Inflow Results

Table 4-7 gives the tidally averaged velocity and salinity profiles along the centerline of the estuary, and the profiles of salinity and tracer dye concentration across the estuary. Compared with the results without a groundwater inflow given in Table 4-1, the effect of the groundwater inflow is to reduce the bottom salinities and increase the surface salinities along most of the estuary, and to change the velocity profiles. The salinity and velocity profile at the mouth of the estuary (I = 18) is changed mostly because there is a greater net outflow of freshwater from the estuary. To get the circulation and transport of constituents correct in estuaries, it is necessary to include the groundwater inflow rate and location.

The lateral slice across the estuary west of the island given in Table 4-7 shows that the groundwater inflow from the extensive shallower areas north of the island changes the salinity profiles outward as far as the main channel. The tracer dye indicates that the initial concentrations in the groundwater are reduced to approximately 6% of their inflow values near the main channel location. The effects of the groundwater inflow on the lateral circulation and the longitudinal variation in the tracer dye concentrations are in the Est_TSC_03 project folder _SPO.dat file.

4.3.2 Additional Suggested Study Examples

Following are some additional suggested study examples.

Est_TSC_03a. Vary the groundwater inflow to determine its influence on circulation and salinity.

Est_TSC_03b. Assume that most of the groundwater enters the deeper channel portion of the estuary from kgrndU = 7 to kgrndL = 14. Determine its effect on salinity profiles, circulation, and tracer dilution for different groundwater inflow rates.

4.4 Coastal Applications

Coastal applications require the specification of tidal boundary conditions at the upcoast location, at the downcoast location, and at the outer boundary at the end of these two locations. One must know how these boundaries are joined to specify their locations properly.

Open boundary elevation conditions applied to the estuary project had only one elevation boundary. Even for this case, it is required that the bathymetry for the first three lines inward from the boundary be identical. This is because some results are interpolated outward from the water body to the open boundary, and errors will occur if there is an upestuary bottom K0(I, J) level which is higher than the downestuary levels.

A more complex problem arises for open boundaries extending offshore into coastal waters. Both the interpolation of constituent concentrations and the transport of momentum through the open boundary require that the next line inward from the boundary for which tidal conditions are specified be part of the tidal boundary condition. An example of how the ends of the open boundaries must be specified and nested is given in Figure 4-4. The coastline is at J = 1, which is a dummy nontransport boundary. The western boundary at I = 2 extends from J = 2 to J = 13, and the column at I = 3 is implied to be part of the same boundary although it is not specified in the input file. Similarly, the eastern boundary at I = 21 extends from J = 2 to J = 13 and the column at I = 20 is implied to be part of that boundary. The outer boundary at J = 13 therefore needs only to extend from I = 4 to I = 19, and the row at J = 12 is implied to be part of that boundary. A completely open water application can also be set up with open tidal boundaries on all four sides. It is difficult to get detailed tidal amplitude and phase shift data for a project application in open water.

The coastal boundaries can also be nested by letting the outer boundary at J = 13 extend from I = 2 to I = 21, and letting the row at J = 12 be the implied boundary. Then the western boundary at I = 2 extends from J = 2 to J = 11, and the eastern boundary at I = 21 also extends from J = 2 to J = 11.

The bathymetry file for the open coastal water project is shown in Table 4-8. The shoreline is from west to east along J = 2. The bottom slopes off uniformly until it hits a shelf at I = 11. The three open boundaries are along I = 2 and I = 21 from J = 2 to 13, and along J = 13 from I = 2 to I = 21. As required for tidal boundaries, the bottom depths are uniform inward from the boundary.

Two projects are examined. The first is for no tidal variations at the boundary. The second is with tides on the three boundaries. For the latter case, there is a tidal lag between the west and east boundaries that is stepped proportionally along the outer boundary. These two projects allow a comparison of the brine plume for a steady-state condition with one generated for time varying tidal conditions.

Both projects have a facility intake of 2 m^3/s at a salinity of approximately 40 ppt that is partially removed so that 1 m^3/s can be used. The remaining 1 m^3/s is discharged as brine with an increase in concentration of 40 ppt above the intake or approximately at a brine concentration of 80 ppt, depending on the amount of recirculation between the discharge and the intake. There is a small increase in temperature across the facility of 5° C.

4.4.1   Coastal_Brine_Discharge_01 Project

The input data for this project is shown in Table 4-9. It has three elevation boundaries each at a mean elevation of 0.20 m, but with no tide. The tidal boundaries are laid out with the outer northern boundary nested between the western and eastern boundaries. Detailed results of the simulation are presented in its example application folder.

Table 4-10 shows the south–north slice of velocity and salinity profiles from the shoreline through the outfall. The salinity profiles show the plume developing on the bottom J = 7 and extending downslope. The velocity profiles show the plume moving downslope along the bottom. This movement generates a reverse current in the upper layers extending from the surface to the top of the plume.

Table 4-11 shows the distribution of the south–north velocity component, and the bottom salinity distribution. The maximum bottom concentration is near the point of discharge, and the plume spreads northward and laterally from that point. The velocity component shows that the maximum bottom velocity is at the point of discharge and it decreases northward and laterally from that location. There is some asymmetry in the velocity field due to the effects of the intake withdrawal.

4.4.2   Coastal_Brine_Discharge_02 Project

The second coastal project applies time varying tides to all three open boundaries. This requires having an overall time lag between the western and eastern boundary, and interpolating that time lag incrementally along the northern boundary.

The input data file for this project is lengthy and therefore not reproduced as a separate table, but it can be found in the project folder. A summary table of the tidal boundary conditions is given in Table 4-12. The time lag between the west and east boundary is 0.016 hr. It is stepped along the northern boundary by 0.004-hr increments for each group of three cells giving five separate tidal boundaries along the northern boundary. The minimum number of cells that could be used is two for each step. The tidal mean elevation and amplitude is 0.20 m with a tidal period of 12.45 hr.

The tidally averaged U-velocity component and bottom salinity distribution are shown in Table 4-13. The plume with lagged tides along all the boundaries shows, in comparison with the steady-state results in Table 4-11, that bottom concentrations are slightly higher. The plume also tends to be stretched toward the eastern boundary in the direction of the net current resulting from the tide lag. The spatial distributions of velocity and salinity for every 1/6 of the last two tidal periods (tmeout = 12.45/6) is given in the project SPO file. It shows the current reversing at different times along the coast.

Because a surface and bottom output is specified, the plotting file ending with _02__PLT_BOT.dat gives for each I and J (as x and y) location the U and V velocity components, temperature, salinity, and the arbitrary constituent concentration. The arbitrary constituent is not used in the present project. The results are given in the file for each 1/6 of a tidal cycle. The file can be used in a plotting routine to see the velocity field and the salinity distribution at each 1/6 of a tidal cycle through the last two tidal periods. The last set of data in the _02_PLT_BOT.dat file gives the tidally averaged velocity components and constituents at each I and J for plotting. They are shown in Figure 4-5 for comparison with the values given in Table 4-11.

4.4.3   Estimating Tidal Time Lags Along Open Boundaries

One problem with open coastal boundaries is determining the tidal time lag along them. The time lag can sometimes be estimated from the tidal or gravity wave speed, which is a function of water depth (Defant 1958) as follows:

$$Vts = (gH)^{1/2} \tag{4.1}$$

where Vts = gravity wave speed, m/s; g = gravitational acceleration, 9.78 m/s^2; and H = water depth, m.

The time lag along a boundary becomes:

$$t_{lag} = L_{bnd}/(gH)^{1/2} \tag{4.2}$$

where t_{lag} = time lag in seconds, and L_{bnd} = distance along the boundary, in meters.

For the example project Coastal_Brine_02, the average depth is approximately 5 m, giving a gravity wave speed of 7 m/s or 2.5×10^4 m/hr. The distance between the western and eastern boundary is 2000 m, suggesting that the time lag between them should be 0.08 hr, or approximately five times the value used in the application.

The problem is more complicated for the bathymetry used in the application. The bottom slopes offshore, making the tidal wave slower near the beach than it is offshore. The tidal wave propagates at an angle across the computational grid.

The use of the gravity wave speed to determine tide lags along a beach is very approximate. Bottom friction, reflection, and refraction can distort the wave.

4.4.4   Additional Suggested Study Examples

The abbreviations CBD_01 and CBD_02 are used to identify the base project that applies to the following suggested study examples.

CBD_01a. The results of CBD_01 indicate that it may be useful to look at greater detail using a bathymetry with more cells (i.e., increase IM and JM) using the same dx, dy, and dz. Using the Bathymetrysetup.exe routine, generate a new grid. Using the INTROGLLVHT Input File.exe routine, generate a new input file and set of output files. Remember to nest the boundaries.

CBD_02a. Turn off the facility intake and discharge to determine what the time varying bottom tidal currents and tidally averaged bottom current look like due to the tidal boundary conditions alone. Is there a net flow in one direction resulting from the time lags between and along the boundaries? Does changing the time lag between the west and east boundaries to the value indicated in Section 4.3.3 make any difference?

CBD_02b. It may be necessary for navigational reasons and protection against wave action to construct the facility intake pipe on top of the discharge pipe where it will be as deep as the discharge pipe but not extending out as far. Assess the effects of different intake locations on recirculation to the facility and on the size of the salinity footprint. Note that project CBD_02 does not have the discharge linked to the intake.

CBD_02c. Increase the tidal amplitude, as well as the mean tide level, along all of the tidal boundaries to assess the effects of tide on the plume size and bottom currents.

CBD_02d. Place a wind parallel to the shore line (the Wx component) and determine its effect on surface currents, bottom currents, plume size, and orientation.

CBD_02e. With the facility turned off, set up a tidal boundary condition with time lags varying along all three boundaries to represent the effect of the tide propagating diagonally across the grid. Assume that the tide hits the northwestern corner of the grid first, and is lagged across the other boundaries from that location. How much difference does this approximation make in the tidally averaged currents?

CBD_02f. Using the conditions established in CBD_02b, turn the facility back on and see if the tidal boundary condition in that application has any influence on the bottom salinity plume distribution.

CBD_02g. Try an open water application with open tidal boundaries on all four sides. Set the bathymetry up for a uniform depth over a square-shaped portion of the ocean. Place a wind with only a Wx component, then a wind with only a Wy component. Try a phase shift between two ends, and stepped along the other sides.

General. The results in Figure 4-5 show an increase in salinity above the original 40 ppt background of 2 ppt in the immediate vicinity of the discharge with a footprint extending out to an increase of 0.25 ppt (250 ppm) above background over an area of 4×10^5 m^2 (40 hectares or approximately 100 acres). The question is, will this have an impact on benthos and other organisms that depend on them for food in the area?

Assuming different geographic locations around the world, assess the environmental impact of this footprint. Try using the Web to search for information to use in the assessment.

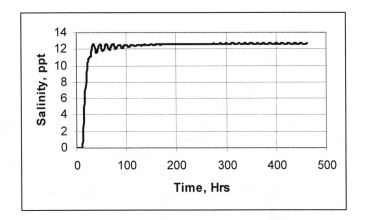

Figure 4-1. Time series of salinity at the head of the estuary showing approach to stationary-state conditions in approximately 100 to 200 hours.

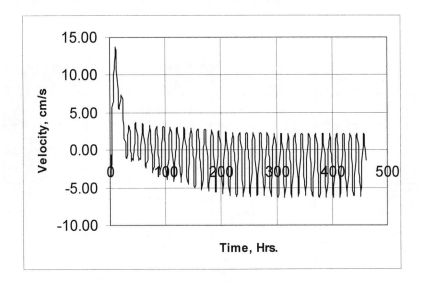

Figure 4-2. Time series of U-velocity component near intake showing approach to stationary state conditions in approximately 300 hours.

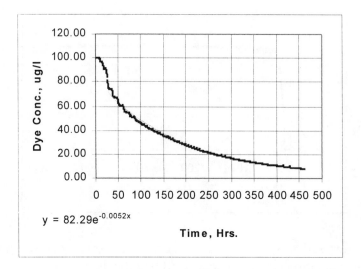

$$y = 82.29e^{-0.0052x}$$

Figure 4-3. Time series of dye concentration at I = 8, J = 2, K = 2 showing decrease over time. Also, fit of exponential die-away giving a half-life or flushing time at that location of 5.5 days.

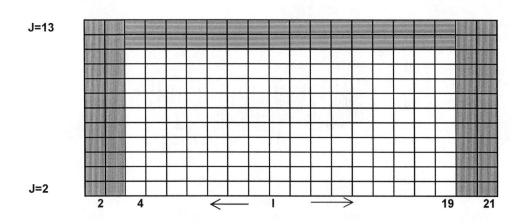

Figure 4-4. Definition diagram of nesting of adjoining open boundaries for open coastal case.

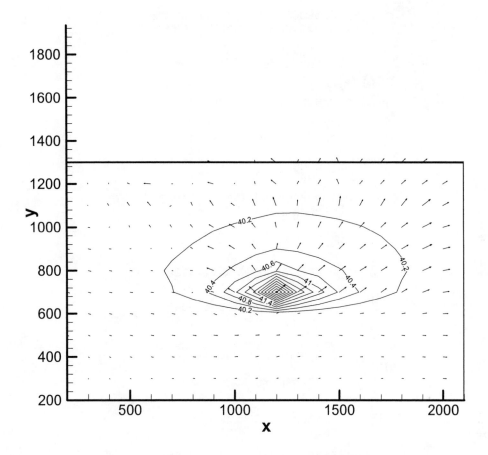

Figure 4-5. Tidally averaged bottom velocities and salinity distribution from Coastal_Brine_Discharge_02_PLT_BOT.dat.

Table 4-1. Profiles of tidally averaged salinity and velocity as a slice along J = 4.

West-East Constituent profiles for arvsaln at j= 4

2.	3.	4.	5.	6.	7.	8.	9.	10.	11.	12.	13.	14.	15.	16.	17.	18.	19.	20.
6.11	7.50	8.30	8.76	9.12	9.39	9.64	9.80	9.86	9.86	9.81	9.78	9.69	9.61	9.56	9.51	9.54	10.00	10.00
9.54	9.71	9.83	9.91	9.97	10.06	10.11	10.21	10.23	10.25	10.24	10.22	10.11	10.07	10.04	10.03	9.97	10.00	10.00
11.28	11.14	11.05	11.04	11.10	11.06	11.09	11.00	11.03	11.08	11.08	10.97	10.89	10.85	10.86	10.88	10.77	12.00	12.00
12.11	12.30	12.48	12.43	12.35	12.36	12.34	12.39	12.38	12.41	12.23	12.31	12.42	12.44	12.38	12.32	12.31	15.00	15.00
0.00	0.00	13.64	13.84	14.06	14.01	14.07	14.20	14.36	14.30	14.40	14.58	14.56	14.62	14.75	14.75	14.61	18.00	18.00
0.00	0.00	0.00	0.00	15.11	15.31	15.57	15.89	16.35	16.42	16.73	17.05	17.10	17.31	17.29	17.40	17.83	20.00	20.00
0.00	0.00	0.00	0.00	0.00	0.00	0.00	0.00	17.02	17.38	17.74	18.51	18.67	19.00	19.07	19.24	19.49	20.00	20.00
0.00	0.00	0.00	0.00	0.00	0.00	0.00	0.00	0.00	0.00	0.00	18.91	19.21	19.48	19.57	19.73	19.91	20.00	20.00
0.00	0.00	0.00	0.00	0.00	0.00	0.00	0.00	0.00	0.00	0.00	0.00	0.00	19.57	19.77	19.84	19.97	0.00	0.00

West-East Arv U-velocity profiles, cm/sec at j= 4

2.	3.	4.	5.	6.	7.	8.	9.	10.	11.	12.	13.	14.	15.	16.	17.	18.	19.	20.
1.74	1.39	1.27	1.18	1.03	0.86	0.37	-0.29	-1.06	-1.00	-1.54	-1.81	-1.70	-2.21	-2.52	-2.77	-0.53	2.35	2.35
-0.96	-0.61	-0.12	-0.22	-0.35	-0.04	-0.05	-0.01	0.17	0.48	0.37	-0.18	-0.04	-0.49	-0.77	-1.07	-0.38	2.35	2.35
-0.62	-0.33	-0.49	-0.65	-0.09	-0.04	-1.03	1.26	1.20	1.45	2.52	2.11	1.81	1.46	1.17	0.83	0.50	2.43	2.43
-1.24	-2.10	-0.01	-0.17	-0.18	0.50	1.22	1.34	1.56	3.40	2.68	2.25	3.94	4.07	3.92	2.62	4.34	3.16	3.16
0.00	0.00	-1.58	-2.53	-0.69	0.21	-1.94	-1.24	0.45	0.59	0.45	-0.35	2.65	2.56	3.25	4.49	3.95	3.30	3.30
0.00	0.00	0.00	0.00	-1.64	-1.66	-2.47	-2.39	-1.81	-2.52	-1.75	-1.93	-1.04	0.49	-0.43	-1.35	-2.43	2.30	2.30
0.00	0.00	0.00	0.00	0.00	-2.23	0.00	0.00	-2.65	-2.89	-2.47	-2.96	-2.99	-2.82	-3.09	-2.70	-3.33	2.34	2.34
0.00	0.00	0.00	0.00	0.00	0.00	0.00	0.00	0.00	0.00	0.00	0.00	-2.18	-2.97	-3.49	-2.42	-3.72	2.31	2.31
0.00	0.00	0.00	0.00	0.00	0.00	0.00	0.00	0.00	0.00	0.00	0.00	0.00	0.00	-2.89	-1.97	-3.40	2.03	2.03
0.00	0.00	0.00	0.00	0.00	0.00	0.00	0.00	0.00	0.00	0.00	0.00	0.00	0.00	0.00	0.00	0.00	0.	0.

Table 4-2. Tidally averaged surface and profile constituent distributions for a facility discharge with an increase of 100.0 ug/l of virtual dye between the intake and discharge, as well as west to east surface velocities.

Surface Constituent Arv for   const

	3.	4.	5.	6.	7.	8.	9.	10.	11.	12.	13.	14.	15.	16.	17.	18.	19.	20.	const
	0.00	0.00	0.00	0.00	0.00	0.00	0.00	0.00	0.00	0.00	0.00	0.00	0.00	0.11	0.12	0.09	0.00	0.00	0.00
	0.00	0.00	0.00	0.00	0.00	0.00	0.00	0.00	0.00	0.00	0.26	0.17	0.12	0.12	0.09	0.07	0.00	0.00	0.00
	0.00	0.00	0.00	0.00	0.00	0.00	0.00	0.00	0.59	0.49	0.30	0.21	0.14	0.13	0.09	0.05	0.00	0.00	0.00
	0.00	0.00	0.00	0.00	0.00	0.00	1.00	1.04	0.90	1.09	0.55	0.23	0.17	0.08	0.06	0.03	0.00	0.00	0.00
	0.28	0.32	0.39	0.48	0.65	0.98	1.23	1.92	2.01	0.00	0.00	0.29	0.12	0.11	0.05	0.04	0.00	0.00	0.00
	0.31	0.35	0.39	0.46	0.64	1.25	2.50	6.75	4.15	0.00	0.00	0.19	0.19	0.11	0.07	0.05	0.00	0.00	0.00
	0.00	0.00	0.00	0.00	0.73	1.05	2.50	6.00	0.00	0.00	0.00	0.00	0.15	0.09	0.08	0.07	0.00	0.00	0.00

West-East Constituent Profiles for arvconstat j=   4

j=	2.	3.	4.	5.	6.	7.	8.	9.	10.	11.	12.	13.	14.	15.	16.	17.	18.	19.	20.
	0.22	0.28	0.32	0.39	0.48	0.65	0.98	1.23	1.92	2.01	1.09	0.55	0.29	0.19	0.11	0.07	0.04	0.00	0.00
	0.28	0.38	0.43	0.50	0.60	0.76	0.97	1.15	1.53	1.79	1.21	0.74	0.42	0.26	0.17	0.10	0.06	0.00	0.00
	0.35	0.43	0.47	0.53	0.61	0.72	0.87	1.01	1.23	1.45	1.19	0.92	0.59	0.43	0.31	0.21	0.11	0.00	0.00
	0.44	0.46	0.48	0.52	0.58	0.66	0.76	0.86	0.97	1.13	1.06	0.88	0.69	0.59	0.48	0.36	0.19	0.00	0.00
	0.00	0.00	0.51	0.53	0.54	0.60	0.65	0.70	0.74	0.81	0.75	0.66	0.57	0.49	0.41	0.32	0.23	0.00	0.00
	0.00	0.00	0.00	0.00	0.50	0.51	0.50	0.49	0.45	0.46	0.40	0.32	0.28	0.22	0.19	0.15	0.08	0.00	0.00
	0.00	0.00	0.00	0.00	0.00	0.00	0.00	0.00	0.34	0.29	0.24	0.13	0.11	0.07	0.06	0.04	0.02	0.00	0.00
	0.00	0.00	0.00	0.00	0.00	0.00	0.00	0.00	0.00	0.00	0.00	0.08	0.05	0.03	0.02	0.01	0.00	0.00	0.00
	0.00	0.00	0.00	0.00	0.00	0.00	0.00	0.00	0.00	0.00	0.00	0.00	0.00	0.02	0.01	0.01	0.00	0.00	0.00
	0.00	0.00	0.00	0.00	0.00	0.00	0.00	0.00	0.00	0.00	0.00	0.00	0.00	0.00	0.00	0.00	0.00	0.00	0.00

Surface West-East Arv U-surface velocities, cm/sec

	2.	3.	4.	5.	6.	7.	8.	9.	10.	11.	12.	13.	14.	15.	16.	17.	18.	19.	20.
	0.00	0.00	0.00	0.00	0.00	0.00	0.00	0.00	0.00	0.00	0.00	0.00	0.00	0.00	-0.63	0.00	0.00	0.00	0.00
	0.00	0.00	0.00	0.00	0.00	0.00	0.00	0.00	0.00	0.00	0.00	0.00	0.00	-0.84	-0.66	0.53	0.00	0.00	0.00
	0.00	0.00	0.00	0.00	0.00	0.00	0.00	-0.95	-1.36	-0.71	-0.73	-0.09	-0.65	-0.64	-0.63	1.02	0.00	0.00	0.00
	0.00	0.00	0.00	0.00	0.00	0.00	0.00	-0.29	-1.06	-1.50	-2.21	-0.54	-0.19	0.61	0.80	-0.89	-2.70	0.00	0.00
	0.00	1.39	1.27	1.18	1.03	0.86	0.37	-0.93	0.94	-1.00	-1.54	-2.99	-2.87	0.00	0.00	-3.78	-0.53	1.37	1.37
	1.74	0.48	0.93	1.30	1.63	1.38	0.77	0.00	0.00	0.00	0.00	-1.81	-1.70	-2.97	-2.81	-3.77	0.41	2.35	2.35
	0.00	0.00	0.00	0.00	0.00	0.14	0.00	0.00	0.00	0.00	0.00	0.00	-1.60	-2.21	-2.52	-2.50	0.00	4.60	4.60
													0.00	-2.15	-2.44	-0.47		0.00	0.00
														0.00	-0.76				

Table 4-3. Input file for project EST_TSC_02 illustrating setup to compute residence and flushing times.

```
Est_TSC_02
 $1.nwqm                  1
 $2.Inflow Conditions
 $ninflows               2
 $qinflow,iinflow,jinflow,kinflow
 $intake,inintake,jintake,kintake
 $temp,saln,const,cbod,on,nh3,do(d),no3,op,po4,phyt
      5.00      2      4      2
      0      0      0      0
      0.00      0.00      0.00
      1.0     10      3      2
      1      8      2      2
      0.00      0.00  0.00
 $.4 Outflow Conditions
 $noutflows              1
 $qoutflow,ioutflow,joutflow,koutflow
      1.0      8      2      2
 $4.Elevation Boundary Conditions
 $nelevation  kts        1              2
 $iewest,ieeast,jesouth,jenorth
 zmean,zamp,tmelag,tideper
 k,temp,saln,const,cbod,on,nh3,do(d),no3,op,po4,phyt for k=2,km-2
     20     20      3      5
      0.40      0.40      0.00     12.45
      2      0.00     10.00      0.00
      3      0.00     10.00      0.00
      4      0.00     10.00      0.00
      5      0.00     12.00      0.00
      6      0.00     15.00      0.00
      7      0.00     18.00      0.00
      8      0.00     20.00      0.00
      9      0.00     20.00      0.00
     10      0.00     20.00      0.00
     11      0.00     20.00      0.00
 $5.Initialize Water Quality Profiles
 $ninitial               1
 k,temp,saln,const,cbod,on,nh3,do(d),no3,op,po4,phyt for k=2,km-2
      2      0.00      0.00    100.00
      3      0.00      0.00    100.00
      4      0.00      0.00    100.00
      5      0.00      0.00    100.00
      6      0.00      0.00    100.00
      7      0.00      0.00    100.00
      8      0.00      0.00    100.00
      9      0.00      0.00    100.00
     10      0.00      0.00    100.00
     11      0.00      0.00    100.00
```

Table 4-3 (continued)

```
$6.External Parameters
 Chezy, Wx,Wy,CSHE, TEQ, Rdecay, Lat.
    35.00      0.00     0.00     25.00      0.00      0.00     30.3
 $7.Output Profiles
 $nprofiles              1
 $ipwest,ipest,jpsouth,jpnorth
 u-vel, v-vel, w-vel
 $nconstituents
 I-const(1),I-const(2), I-const(3), etc
      2      20      4       4
      1       0      1
      0
    $8.Output Surfaces
 $nsurfaces   $nconstituents             1           0
 $U-vel   V-vel   0.000000          0.000000
 $I-const(1), I-const(2), I-const(3), etc.
  $9.Output Time Series
 $ntimser              3
 $nconst, iconst, jconst,kconst
      3       8      2       2
     21      10      4       2
      2       2      4       5
 $10.Simulation time conditions
 $dtm   tmend    120.0000         480.0
 $tmeout  tmeserout    480.0      1.0
 $11.Internal Boundary Locations
 $nintbnd              0
 $ibwest,ibeast,jbsouth,jbnorth,ktop,kbottom
$12.Constituent Averages
 $nconarv       2
 $constarv
3
$13. Groundwater Inflow
 $ngrndwtr              0
```

Table 4-4. Contents of ABControls.dat file for running the EST_TSC_02 project for the computation of residence time.

```
Estuary_Bath.dat
EST_TSC_02_INP.dat
EST_TSC_01_WQM.dat
```

Table 4-5. Profile and surface constituent distributions as converted to residence times in days.

West-East Constituent profiles for arvDays at j=

4

2.	3.	4.	5.	6.	7.	8.	9.	10.	11.	12.	13.	14.	15.	16.	17.	18.	19.	20.
4.76	5.09	5.28	5.36	5.32	5.05	4.46	3.84	3.67	3.53	3.18	2.97	2.70	2.54	2.33	2.07	1.80	0.00	0.00
5.42	5.45	5.42	5.34	5.21	4.97	4.57	4.09	4.02	3.87	3.53	3.45	3.05	2.82	2.64	2.38	2.04	0.00	0.00
5.50	5.45	5.39	5.32	5.23	5.11	4.84	4.44	4.40	4.18	3.92	3.77	3.43	3.19	3.07	2.76	2.35	0.00	0.00
5.43	5.29	5.16	5.12	5.08	4.98	4.88	4.72	4.60	4.26	4.18	3.98	3.70	3.53	3.48	3.22	2.72	0.00	0.00
0.00	0.00	4.91	4.77	4.60	4.48	4.36	4.24	4.22	4.07	3.92	3.75	3.56	3.48	3.39	3.22	2.99	0.00	0.00
0.00	0.00	0.00	0.00	4.31	4.15	3.93	3.69	3.54	3.46	3.28	3.24	3.04	2.97	2.84	2.29	2.45	0.00	0.00
0.00	0.00	0.00	0.00	0.00	0.00	0.00	0.00	3.40	3.19	2.94	2.60	2.56	2.51	2.41	1.95	1.99	0.00	0.00
0.00	0.00	0.00	0.00	0.00	0.00	0.00	0.00	0.00	0.00	0.00	2.47	2.31	2.12	2.10	1.75	1.66	0.00	0.00
0.00	0.00	0.00	0.00	0.00	0.00	0.00	0.00	0.00	0.00	0.00	0.00	0.00	2.03	1.86	0.00	1.46	0.00	0.00
0.00	0.00	0.00	0.00	0.00	0.00	0.00	0.00	0.00	0.00	0.00	0.00	0.00	0.00	0.00	0.00	1.38	0.00	0.00

Surface Constituent Arv for Days

2.	3.	4.	5.	6.	7.	8.	9.	10.	11.	12.	13.	14.	15.	16.	17.	18.	19.	20.
0.00	0.00	0.00	0.00	0.00	0.00	0.00	0.00	0.00	0.00	0.00	0.00	0.00	0.00	3.63	3.44	0.00	0.00	0.00
0.00	0.00	0.00	0.00	0.00	0.00	0.00	0.00	0.00	0.00	0.00	0.00	3.49	3.50	3.38	3.13	0.00	0.00	0.00
0.00	0.00	0.00	0.00	0.00	0.00	0.00	0.00	0.00	0.00	0.00	3.25	3.34	3.44	3.17	2.95	2.98	0.00	0.00
0.00	0.00	0.00	0.00	0.00	0.00	0.00	0.00	0.00	3.29	3.14	3.02	2.93	3.40	3.19	2.64	2.33	0.00	0.00
4.76	5.09	5.28	5.36	5.32	5.05	4.46	3.72	3.47	3.23	3.02	2.97	2.70	2.45	3.00	2.36	2.70	0.00	0.00
0.00	0.00	5.43	5.47	5.46	5.37	4.98	3.84	3.67	3.53	3.18	2.97	2.65	2.54	2.37	2.17	1.80	1.90	0.00
0.00	0.00	0.00	0.00	0.00	5.44	5.27	4.44	4.24	4.14	0.00	0.00	2.62	2.52	2.49	2.41	2.30	0.00	0.00

Table 4-6. Extension of Est_TSC_01 project input file to include groundwater inflow in project Est_TSC_03.

```
$7.Output Profiles
 $nprofiles                    2
 $ipwest,ipest,jpsouth,jpnorth
 u-vel, v-vel, w-vel
 $nconstituents
 I-const(1),I-const(2), I-const(3), etc
    2      20      4       4
    1      0       0
    2
    2       3
   14      14      2      10
    0       1      0
    2
    2       3
   $8.Output Surfaces
 $nsurfaces   $nconstituents              0            0
 $U-vel  V-vel   0.000000        0.000000
 $I-const(1), I-const(2), I-const(3), etc.
  $9.Output Time Series
 $ntimser              3
 $nconst, iconst, jconst,kconst
    3       8      2       2
   21      10      4       2
    2       2      4       5
$10.Simulation time conditions
$dtm  tmend   120.0000        480.0
$tmeout  tmeserout    6.22     1.0
$11.Internal Boundary Locations
$nintbnd            0
$ibwest,ibeast,jbsouth,jbnorth,ktop,kbottom
$12.Constituent Maxima
 $nconarv      2
 $constarv
   2  3
$13. Groundwater Inflow
$ngrndwtr          1
$qgrndwtr, kgrndU kgrndL
        20      3     4
$temp saln const
0.0  0.0 1000.0
```

Table 4-7. Estuary with groundwater inflow, Est_TSC_03 project.

West-East Arv u-velocity profiles, cm/sec at j= 4

2.	3.	4.	5.	6.	7.	8.	9.	10.	11.	12.	13.	14.	15.	16.	17.	18.	19.	20.
1.67	2.00	2.46	1.95	1.53	1.20	0.51	0.46	-0.18	-0.20	-0.55	-0.55	-0.46	-1.05	-1.76	-1.81	0.60	2.59	2.59
-0.51	-0.63	-0.02	-0.32	-0.05	0.25	0.47	0.04	-0.03	0.62	0.72	0.98	0.71	-0.18	-0.23	-0.47	0.56	2.57	2.57
-0.28	-0.29	-1.28	-1.27	-0.07	0.41	1.69	1.50	0.71	0.70	2.21	2.07	1.55	1.36	1.34	1.27	1.21	2.63	2.63
-1.24	-2.27	-0.47	-0.26	-0.49	0.40	1.16	0.99	1.19	2.69	1.61	1.44	3.04	3.06	3.21	3.38	4.09	3.18	3.19
0.00	0.00	-1.65	-2.69	-0.89	-2.15	-2.48	-1.81	-0.22	-0.50	-0.69	1.17	1.44	1.82	2.64	3.49	3.15	3.18	3.18
0.00	0.00	0.00	0.00	-1.72	-2.30	-2.55	-2.45	-2.71	-2.85	-2.26	-1.05	-1.76	-0.22	-1.14	-1.88	-1.94	2.38	2.89
0.00	0.00	0.00	0.00	0.00	0.00	0.00	0.00	0.00	-3.03	-2.54	-2.04	-3.03	-2.82	-3.21	-2.81	-2.54	2.43	2.43
0.00	0.00	0.00	0.00	0.00	0.00	0.00	0.00	0.00	0.00	0.00	-2.92	-2.08	-2.96	-3.48	-2.41	-3.38	2.40	2.40
0.00	0.00	0.00	0.00	0.00	0.00	0.00	0.00	0.00	0.00	0.00	0.00	0.00	0.00	-2.89	-1.96	-3.72	2.10	2.10
0.00	0.00	0.00	0.00	0.00	0.00	0.00	0.00	0.00	0.00	0.00	0.00	0.00	0.00	0.00	0.00	-3.37	0.00	0.00
0.00	0.00	0.00	0.00	0.00	0.00	0.00	0.00	0.00	0.00	0.00	0.00	0.00	0.00	0.00	0.00	0.00	0.00	0.00

West-East Constituent profiles for arvsaln at j= 4

2.	3.	4.	5.	6.	7.	8.	9.	10.	11.	12.	13.	14.	15.	16.	17.	18.	19.	20.
5.90	7.33	8.09	8.26	8.52	8.79	9.05	9.24	9.34	9.40	9.40	9.43	9.40	9.39	9.40	9.36	9.39	10.00	10.00
7.33	9.06	9.20	9.33	9.38	9.48	9.55	9.61	9.67	9.74	9.77	9.79	9.77	9.80	9.82	9.80	9.78	10.00	10.00
8.98	10.69	10.59	10.56	10.66	10.63	10.64	10.54	10.53	10.52	10.63	10.54	10.54	10.56	10.58	10.61	10.56	10.00	10.00
10.81	12.13	12.35	12.28	12.17	12.17	12.15	12.23	12.16	12.51	12.03	12.12	12.16	12.26	12.21	12.16	12.19	12.00	12.00
11.92	13.92	13.69	14.15	14.15	14.09	15.75	14.30	14.47	14.41	14.53	14.72	14.64	14.70	14.78	14.78	14.66	15.00	15.00
0.00	0.00	0.00	0.00	15.25	15.48	0.00	16.06	16.54	16.62	16.92	17.23	17.73	19.07	17.14	17.56	17.94	18.00	18.00
0.00	0.00	0.00	0.00	0.00	0.00	0.00	0.00	17.16	17.51	17.85	18.96	18.76	19.51	19.60	19.32	19.51	20.00	20.00
0.00	0.00	0.00	0.00	0.00	0.00	0.00	0.00	0.00	0.00	0.00	0.00	19.25	19.59	19.78	19.75	19.97	20.00	20.00
0.00	0.00	0.00	0.00	0.00	0.00	0.00	0.00	0.00	0.00	0.00	0.00	0.00	0.00	0.00	19.85	19.97	20.00	20.00
0.00	0.00	0.00	0.00	0.00	0.00	0.00	0.00	0.00	0.00	0.00	0.00	0.00	0.00	0.00	0.00	0.00	20.00	20.00

South-North Constituent Profiles for Arvsaln at i= 14

2.	3.	4.	5.	6.	7.	8.	9.	10.
0.00	9.43	9.40	9.32	9.17	8.97	8.88	0.00	0.00
0.00	9.74	9.77	9.70	9.66	9.44	9.39	0.00	0.00
0.00	10.43	10.54	10.56	10.45	10.53	10.40	0.00	0.00
0.00	12.46	12.26	12.20	12.35	12.13	11.86	0.00	0.00
0.00	14.46	14.88	14.88	14.44	14.44	0.00	0.00	0.00
0.00	0.00	16.97	16.97	0.00	0.00	0.00	0.00	0.00
0.00	0.00	0.00	0.00	0.00	0.00	0.00	0.00	0.00

South-North Constituent Profiles for Arvconstat i= 14

2.	3.	4.	5.	6.	7.	8.	9.	10.
0.00	42.30	62.75	84.92	148.14	194.58	204.46	0.00	0.00
0.00	48.85	67.45	85.18	127.56	166.04	180.42	0.00	0.00
0.00	56.55	74.17	88.33	114.71	131.65	141.18	0.00	0.00
0.00	53.09	68.50	79.29	89.93	97.25	103.24	0.00	0.00
0.00	42.87	49.38	50.69	58.86	0.00	0.00	0.00	0.00
0.00	0.00	23.01	26.06	0.00	0.00	0.00	0.00	0.00
0.00	0.00	8.89	0.00	0.00	0.00	0.00	0.00	0.00
0.00	0.00	4.46	0.00	0.00	0.00	0.00	0.00	0.00

Table 4-8. Bathymetry data file for coastal water project.

```
Coastal_01
22   IM
14   jm
24   km
100    dx
100    dy
1   dz
       2   3   4   5   6   7   8   9  10  11  12  13  14  15  16  17  18 19 20 21
13 11 11 11 11 11 11 11 11 11 11 11 11 11 11 11 11 11 11 11
12 11 11 11 11 11 11 11 11 11 11 11 11 11 11 11 11 11 11 11
11 11 11 11 11 11 11 11 11 11 11 11 11 11 11 11 11 11 11 11
10 10 10 10 10 10 10 10 10 10 10 10 10 10 10 10 10 10 10 10
9   9   9   9   9   9   9   9   9   9   9   9   9   9   9   9   9  9  9  9
9   9
8   8   8   8   8   8   8   8   8   8   8   8   8   8   8   8  8  8  8
8   8
7   7   7   7   7   7   7   7   7   7   7   7   7   7   7   7  7  7  7
7   7
6   6   6   6   6   6   6   6   6   6   6   6   6   6   6   6  6  6  6
6   6
5   5   5   5   5   5   5   5   5   5   5   5   5   5   5   5  5  5  5
5   5
4   4   4   4   4   4   4   4   4   4   4   4   4   4   4   4  4  4  4
4   4
3   3   3   3   3   3   3   3   3   3   3   3   3   3   3   3  3  3  3
3   3
2   2   2   2   2   2   2   2   2   2   2   2   2   2   2   2  2  2  2
2   2
-999    geomdelim
```

Table 4-9. Input data for Coastal_Brine_Discharge_01 example application project.

```
Coastal_Brine_Discharge_01
 $1.nwqm              1
 $2.Inflow Conditions
 $ninflows            1
 $qinflow,iinflow,jinflow,kinflow
 $intake,inintake,jintake,kintake
 $temp,saln,const,cbod,on,nh3,do(d),no3,op,po4,phyt
     1.0  12    7    8
      1      12      5     4
    5.00    40.00      0.00
 $.4 Outflow Conditions
 $noutflows               1
 $qoutflow,ioutflow,joutflow,koutflow
  2.0    12   5      4
 $4.Elevation Boundary Conditions
 $nelevation  kts            3           2
 $iewest,ieeast,jesouth,jenorth
 zmean,zamp,tmelag,tideper
 k,temp,saln,const,cbod,on,nh3,do(d),no3,op,po4,phyt for k=2,km-2
     2      2      2     13
     0.2      0.0      0.00        0.0
     2    20.00    40.00       0.00
     3    20.00    40.00       0.00
     4    20.00    40.00       0.00
     5    20.00    40.00       0.00
     6    20.00    40.00       0.00
     7    20.00    40.00       0.00
     8    20.00    40.00       0.00
     9    20.00    40.00       0.00
    10    20.00    40.00       0.00
    11    20.00    40.00       0.00
    12    20.00    40.00       0.00
    13    20.00    40.00       0.00
    14    20.00    40.00       0.00
    15    20.00    40.00       0.00
    16    20.00    40.00       0.00
    17    20.00    40.00       0.00
    18    20.00    40.00       0.00
    19    20.00    40.00       0.00
    20    20.00    40.00       0.00
    21    20.00    40.00       0.00
    22    20.00    40.00       0.00
   21     21      2     13
     0.2      0.0      0.0            0.0
     2    20.00    40.00       0.00
     3    20.00    40.00       0.00
     4    20.00    40.00       0.00
     5    20.00    40.00       0.00
     6    20.00    40.00       0.00
     7    20.00    40.00       0.00
     8    20.00    40.00       0.00
     9    20.00    40.00       0.00
    10    20.00    40.00       0.00
    11    20.00    40.00       0.00
    12    20.00    40.00       0.00
    13    20.00    40.00       0.00
    14    20.00    40.00       0.00
    15    20.00    40.00       0.00
    16    20.00    40.00       0.00
    17    20.00    40.00       0.00
    18    20.00    40.00       0.00
    19    20.00    40.00       0.00
    20    20.00    40.00       0.00
    21    20.00    40.00       0.00
    22    20.00    40.00       0.00
```

Table 4-9 (continued)

```
4    19      13   13
0.2       0.0    0.00     0.0
2     20.00   40.00    0.00
3     20.00   40.00    0.00
4     20.00   40.00    0.00
5     20.00   40.00    0.00
6     20.00   40.00    0.00
7     20.00   40.00    0.00
8     20.00   40.00    0.00
9     20.00   40.00    0.00
10    20.00   40.00    0.00
11    20.00   40.00    0.00
12    20.00   40.00    0.00
13    20.00   40.00    0.00
14    20.00   40.00    0.00
15    20.00   40.00    0.00
16    20.00   40.00    0.00
17    20.00   40.00    0.00
18    20.00   40.00    0.00
19    20.00   40.00    0.00
20    20.00   40.00    0.00
21    20.00   40.00    0.00
22    20.00   40.00    0.00
$5.Initialize Water Quality Profiles
$ninitial            1
k,temp,saln,const,cbod,on,nh3,do(d),no3,op,po4,phyt for k=2,km-2
2     20.00   40.00    0.00
3     20.00   40.00    0.00
4     20.00   40.00    0.00
5     20.00   40.00    0.00
6     20.00   40.00    0.00
7     20.00   40.00    0.00
8     20.00   40.00    0.00
9     20.00   40.00    0.00
10    20.00   40.00    0.00
11    20.00   40.00    0.00
12    20.00   40.00    0.00
13    20.00   40.00    0.00
14    20.00   40.00    0.00
15    20.00   40.00    0.00
16    20.00   40.00    0.00
17    20.00   40.00    0.00
18    20.00   40.00    0.00
19    20.00   40.00    0.00
20    20.00   40.00    0.00
21    20.00   40.00    0.00
22    20.00   40.00    0.00
```

Table 4-9 (continued)

```
$6.External Parameters
Chezy, Wx,Wy,CSHE, TEQ, Rdecay, Lat.
    35.00      0.00      0.00      0.00      0.00      0.00  15.5
$7.Output Profiles
$nprofiles              2
$ipwest,ipest,jpsouth,jpnorth
u-vel, v-vel, w-vel
$nconstituents
I-const(1),I-const(2), I-const(3), etc
    12     12     2     12
     0      1     0
     2
     1      2
     2     20     6     6
     1      0     0
     2
     1      2
$8.Output Surfaces
$nsurfaces  $nconstituents              2              1
$U-vel   v-vel  1  1
$I-const(1), I-const(2), I-const(3), etc.
     2
$9.Output Time Series
$ntimser              2
$nconst, iconst, jconst,kconst
      2 12  5  4
        2 12  7  8
$10.Simulation time conditions
$dtm  tmend   120.00000          240.00000
$tmeout  tmeserout   240.0    2.0
$11.Internal Boundary Locations
$nintbnd            0
$ibwest,ibeast,jbsouth,jbnorth,ktop,kbottom
$12.Constituent Averages
$nconarv        0
$constarv
$13. Groundwater Inflow
$ngrndwtr              0
```

Table 4-10. South-north slice through the discharge showing velocity and salinity profiles.

South-North v-velocity profiles, cm/sec at i= 12

2.	3.	4.	5.	6.	7.	8.	9.	10.	11.	12.
0.30	0.66	1.10	1.16	1.09	-0.84	-2.13	-1.97	-1.64	-1.54	0.01
0.22	0.55	1.04	1.15	1.17	-0.89	-2.22	-2.05	-1.82	-2.18	0.08
0.00	0.39	0.89	1.12	1.27	-0.87	-2.26	-2.13	-2.00	-2.71	0.38
0.00	0.00	0.63	0.88	1.40	-0.79	-2.25	-2.12	-2.02	-2.95	0.98
0.00	0.00	0.00	0.66	1.56	-0.56	-2.04	-1.61	-1.33	-2.48	1.69
0.00	0.00	0.00	0.00	1.43	0.25	-0.60	-0.33	-0.33	-1.37	2.37
0.00	0.00	0.00	0.00	0.00	5.33	3.69	1.93	1.28	0.29	2.92
0.00	0.00	0.00	0.00	0.00	0.00	4.13	4.49	3.38	2.33	3.38
0.00	0.00	0.00	0.00	0.00	0.00	0.00	3.77	4.88	4.56	3.79
0.00	0.00	0.00	0.00	0.00	0.00	0.00	0.00	3.92	6.33	4.08
0.00	0.00	0.00	0.00	0.00	0.00	0.00	0.00	0.00	6.35	3.69
0.00	0.00	0.00	0.00	0.00	0.00	0.00	0.00	0.00	0.00	0.00

South-North Constituent Profiles forsaln at i= 12

2.	3.	4.	5.	6.	7.	8.	9.	10.	11.	12.
40.00	40.00	40.00	40.00	40.00	40.00	40.00	40.00	40.00	40.00	40.00
40.00	40.00	40.00	40.00	40.00	40.00	40.00	40.00	40.00	40.00	40.00
0.00	0.00	40.00	40.00	40.00	40.00	40.00	40.00	40.00	40.00	40.00
0.00	0.00	0.00	40.00	40.01	40.01	40.05	40.01	40.01	40.01	40.00
0.00	0.00	0.00	0.00	40.05	40.05	40.74	40.06	40.02	40.04	40.00
0.00	0.00	0.00	0.00	0.00	42.46	40.74	40.25	40.09	40.08	40.00
0.00	0.00	0.00	0.00	0.00	0.00	0.00	40.41	40.22	40.15	40.00
0.00	0.00	0.00	0.00	0.00	0.00	0.00	0.00	40.29	40.19	40.00
0.00	0.00	0.00	0.00	0.00	0.00	0.00	0.00	0.00	40.19	40.00
0.00	0.00	0.00	0.00	0.00	0.00	0.00	0.00	0.00	0.00	0.00

Table 4-11. Bottom south-north velocity component and salinity distribution.

Bottom South-North v-bottom velocities, cm/sec

```
 0.00  0.00  1.97  2.32  2.62  2.87  3.10  3.30  3.47  3.59  3.69  3.70  3.82  3.80  3.61  3.34  3.00  2.46  0.00  0.00
 1.28  1.28  1.97  2.32  2.62  2.87  3.10  3.30  3.47  3.59  3.69  3.70  3.82  3.80  3.61  3.34  3.00  2.46  1.84  1.81
 1.32  1.32  2.98  3.71  4.25  4.72  5.15  5.55  5.90  6.16  6.35  6.44  6.56  6.43  6.07  5.59  5.02  4.12  2.16  2.08
 0.54  0.54  1.85  2.11  2.36  2.61  2.87  3.18  3.56  3.82  3.92  4.16  4.23  4.00  3.59  3.13  2.84  2.63  1.22  1.13
 0.24  0.25  1.69  1.79  1.89  2.09  2.35  2.73  3.25  3.65  3.77  4.12  4.00  3.60  3.07  2.16  2.34  2.44  1.06  0.97
-0.11 -0.10  1.22  1.46  1.85  2.07  2.27  2.57  3.04  3.74  4.13  4.34  3.93  3.25  2.74  2.42  2.31  2.41  0.73  0.68
-0.23 -0.23 -0.30  0.15  0.90  1.53  2.15  2.75  3.28  3.97  5.33  4.56  3.83  3.39  3.01  2.56  2.04  1.72  0.20  0.19
-0.44 -0.30 -0.38 -0.32 -0.45 -0.57 -0.43 -0.51 -0.40 -0.47  1.43  1.91  1.99  1.69  0.93  0.48  0.43  0.39  0.04  0.04
-0.49 -0.38 -0.42 -0.30 -0.22 -0.20 -0.21 -0.18 -0.15 -0.17  0.66  1.13  1.25  0.99  0.56  0.49  0.39  0.33  0.10  0.10
-0.54 -0.42 -0.44 -0.32 -0.18 -0.08  0.02  0.13 -0.26  0.43  0.39  0.62  0.60  0.46  0.42  0.35  0.32  0.29  0.09  0.09
-0.54 -0.44 -0.37 -0.27 -0.20 -0.09  0.01  0.11  0.21  0.32  0.22  0.40  0.36  0.32  0.34  0.23  0.23  0.25  0.08  0.08
-0.43 -0.43 -0.27 -0.16 -0.08 -0.01  0.06  0.12  0.18  0.21        0.21  0.19  0.17  0.21  0.18        0.18  0.07  0.07
```

Bottom Constituent for saln

```
40.00 40.00 40.00 40.00 40.00 40.00 40.00 40.00 40.00 40.00 40.00 40.00 40.00 40.00 40.00 40.00 40.00 40.00 40.00 40.00
40.00 40.00 40.00 40.00 40.00 40.00 40.00 40.00 40.00 40.00 40.00 40.00 40.00 40.00 40.00 40.00 40.00 40.00 40.00 40.00
40.00 40.00 40.05 40.07 40.10 40.12 40.14 40.16 40.19 40.19 40.18 40.19 40.18 40.17 40.15 40.14 40.12 40.09 40.00 40.00
40.00 40.00 40.07 40.10 40.12 40.15 40.18 40.23 40.29 40.30 40.28 40.28 40.26 40.26 40.23 40.21 40.17 40.14 40.00 40.00
40.00 40.00 40.08 40.11 40.16 40.21 40.27 40.34 40.44 40.41 40.40 40.40 40.35 40.32 40.31 40.26 40.21 40.17 40.00 40.00
40.00 40.00 40.09 40.09 40.21 40.31 40.44 40.56 40.74 40.74 40.70 40.60 40.49 40.42 40.40 40.31 40.26 40.19 40.00 40.00
40.00 40.00 40.00 40.01 40.26 40.42 41.18 42.46 41.51 41.53 41.02 41.02 41.03 40.47 40.41 40.34 40.31 40.14 40.00 40.00
40.00 40.00 40.02 40.03 40.06 40.06 40.08 40.06 40.08 40.05 40.05 40.06 40.02 40.01 40.00 40.00 40.00 40.00 40.00 40.00
40.00 40.00 40.00 40.00 40.00 40.00 40.00 40.00 40.00 40.00 40.00 40.00 40.00 40.00 40.00 40.00 40.00 40.00 40.00 40.00
40.00 40.00 40.00 40.00 40.00 40.00 40.00 40.00 40.00 40.00 40.00 40.00 40.00 40.00 40.00 40.00 40.00 40.00 40.00 40.00
40.00 40.00 40.00 40.00 40.00 40.00 40.00 40.00 40.00 40.00 40.00 40.00 40.00 40.00 40.00 40.00 40.00 40.00 40.00 40.00
40.00 40.00 40.00 40.00 40.00 40.00 40.00 40.00 40.00 40.00 40.00 40.00 40.00 40.00 40.00 40.00 40.00 40.00 40.00 40.00
```

Table 4-12. Open boundary specifications for project Coastal_Brine_TSC_02.

Tidal Boundary	Iewest	Ieeast	Iesouth	ienorth	Tmelag, hrs
1	2	2	2	13	0.0
2	21	21	2	13	0.016
3	4	6	13	13	0.0
4	7	9	13	13	0.004
5	10	12	13	13	0.008
6	13	15	13	13	0.012
7	16	19	13	13	0.016

Table 4-13. Tidally averaged bottom U-velocity component and salinity distribution for Coastal_Brine_02 project with tides along all boundaries.

```
Bottom West-East Arv U- velocities, cm/sec
-1.78 -1.77 -2.03 -2.24 -0.56 -0.13 -0.74  0.02  0.48  0.00  0.56  1.38  0.91  1.10  2.56  2.50  2.54  2.73  2.20  2.20
-2.12 -2.09 -2.08 -2.30 -0.61 -0.17 -0.78  0.00  0.47  0.00  0.59  1.44  0.97  1.18  2.65  2.59  2.63  2.80  2.33  2.33
-3.00 -3.06 -2.47 -1.97 -1.54 -1.19 -1.01 -0.45 -0.13  0.04  0.69  1.47  1.76  2.21  2.67  3.13  3.49  4.47  2.89  2.89
-2.36 -2.41 -1.95 -1.76 -1.69 -1.51 -1.28 -1.12 -0.85 -0.22  0.88  1.71  2.37  2.76  2.97  3.21  3.61  4.37  3.02  3.02
-1.66 -1.69 -1.48 -1.57 -1.72 -1.79 -1.85 -1.91 -1.72 -0.96  1.50  2.60  3.17  3.41  3.43  3.44  3.54  3.72  2.90  2.90
-0.72 -0.70 -0.46 -1.01 -1.41 -1.86 -2.33 -2.81 -3.02 -2.41  2.93  3.85  4.14  3.99  3.61  3.32  3.00  2.83  2.65  2.65
 0.44  0.52  0.52  0.14 -0.35 -1.04 -1.92 -3.25 -4.74 -5.06  5.73  5.67  4.86  3.91  3.12  2.50  1.90  1.55  2.20  2.20
 1.12  1.31  1.34  1.26  1.19  1.12  1.04  0.99  0.81  0.94  0.61  0.91  0.89  0.96  0.90  0.80  0.59  0.31  1.75  1.76
 1.22  1.35  1.33  1.31  1.29  1.29  1.29  1.29  1.26  1.19  0.80  0.69  0.63  0.59  0.55  0.50  0.38  0.05  1.61  1.61
 1.01  1.13  1.16  1.18  1.19  1.20  1.20  1.18  1.12  0.98  0.73  0.57  0.48  0.44  0.42  0.39  0.32  0.01  1.58  1.58
 0.79  0.89  0.95  1.04  1.04  1.05  1.05  1.01  0.94  0.81  0.65  0.52  0.43  0.36  0.32  0.28  0.23 -0.01  1.49  1.49
 0.56  0.64  0.74  0.82  0.87  0.89  0.89  0.86  0.79  0.69  0.57  0.45  0.36  0.28  0.22  0.16  0.11 -0.08  1.31  1.31
```

```
Bottom Constituent Arv for  saln
40.00 40.00 40.00 40.00 40.00 40.00 40.00 40.00 40.00 40.00 40.00 40.00 40.00 40.00 40.00 40.00 40.00 40.00 40.00 40.00
40.00 40.00 40.00 40.00 40.00 40.00 40.00 40.00 40.00 40.00 40.00 40.00 40.00 40.00 40.00 40.00 40.00 40.00 40.00 40.00
40.00 40.00 40.00 40.00 40.00 40.07 40.10 40.12 40.14 40.15 40.16 40.17 40.25 40.00 40.00 40.00 40.00 40.00 40.00 40.00
40.00 40.00 40.00 40.00 40.00 40.09 40.11 40.15 40.16 40.18 40.20 40.22 40.26 40.34 40.00 40.00 40.00 40.00 40.00 40.00
40.00 40.00 40.00 40.00 40.00 40.11 40.13 40.16 40.18 40.22 40.26 40.33 40.50 40.57 40.75 40.00 40.00 40.00 40.00 40.00
40.00 40.00 40.00 40.00 40.00 40.12 40.12 40.19 40.22 40.28 40.41 40.57 40.75 40.88 41.35 40.06 40.00 40.00 40.00 40.00
40.00 40.00 40.00 40.00 40.00 40.08 40.09 40.17 40.18 40.22 40.27 40.75 41.35 42.48 41.25 40.76 40.46 40.00 40.00 40.00
40.00 40.00 40.00 40.00 40.00 40.05 40.03 40.13 40.11 40.06 40.02 40.88 40.60 41.25 40.43 40.34 40.06 40.00 40.00 40.00
40.00 40.00 40.00 40.00 40.00 40.01 40.01 40.01 40.01 40.00 40.00 40.06 40.03 40.06 40.02 40.00 40.00 40.00 40.00 40.00
40.00 40.00 40.00 40.00 40.00 40.00 40.00 40.00 40.00 40.00 40.00 40.00 40.00 40.00 40.00 40.00 40.00 40.00 40.00 40.00
40.00 40.00 40.00 40.00 40.00 40.00 40.00 40.00 40.00 40.00 40.00 40.00 40.00 40.00 40.00 40.00 40.00 40.00 40.00 40.00
40.00 40.00 40.00 40.00 40.00 40.00 40.00 40.00 40.00 40.00 40.00 40.00 40.00 40.00 40.00 40.00 40.00 40.00 40.00 40.00
```

5. APPLICATION OF THE TEMPERATURE, SALINITY, FIRST-ORDER DECAY CONSTITUENT MODEL TO LAKES AND RESERVOIRS

Lake and reservoir simulations with the INTROGLLVHT modeling have simpler setups and input files than the simulations for the tidal estuary and coastal projects. There are, however, a number of important types of flow conditions occurring in lakes and reservoirs that affect water quality constituent distributions within them. In this chapter, the lake and reservoir cases are set up for the TSC model to show the types of flow patterns produced by different temperature inflows, inflow locations, and surface heat exchange conditions. In later chapters, the lake and reservoir projects will be applied to the DOD and WQDPM water quality models.

The data file for the lake and reservoir projects is given in Table 5-1. The data were obtained by overlaying a U.S.Geological Survey Quad sheet that contains the contoured lake bathymetry with a transparent 400 m by 400 m grid. The depths were then placed at the center of each grid cell.

The reservoir bathymetric map, drawn from the Reservoir_BATH_PLT data file, is shown in Figure 5-1. The reservoir has a main stem running from I = 2, J = 8 to I = 12, J = 9, which is the main inflow. It has a side arm entering at I = 8, J = 2. There is a dam extending across the eastern end of the reservoir at I = 12 from J = 6 to J = 11 for a length of 2400 m, and there are outflows at the dam.

The reservoir has a maximum depth of about 11 m, and a fairly steep bottom along part of the northern side. There are extensive bottom flats between the 2 m and 4 m contour. The reservoir has a surface area of approximately 9.8 x 10^6 m^2 or 980 hectares (2420 acres). This size reservoir could be a small recreational lake or a medium-sized steam electric plant recirculating cooling lake.

5.1 Lake with Cold Water Inflow

The bathymetry in Table 5-1 and Figure 5-1 is first treated as a small recreational lake. The project name for the first application is Res_TSC_01. The input data for this project is shown in Table 5-2.

The main stem inflow at I = 2, J = 8, and K = 2 is 5.0 m^3/s at a temperature of 20° C. The branch inflow at I = 8, J = 2, and K = 2 is 0.50 m^3/s at a temperature of 20° C. The outflow at the dam, I = 12, J = 9, and K = 11, is 5.5 m^3/s and is exactly equal to the sum of the two inflows. In INTROGLLVHT, it is necessary for the sum of the outflows to equal the sum of the inflows because only steady reservoir problems can be run with the model. The reservoir temperature profile is initialized uniformly at 30° C, and the constituent is initialized at 100 ug/l as a virtual dye for the estimate of residence time.

For external parameters, the Chezy bottom friction coefficient is set at 35 m$^{1/2}$/s and there are no surface wind shear components. The coefficient of surface heat exchange is set at 25 watt/m^2/°C and the equilibrium temperature of surface heat exchange is set at 35° C such that there will be a tendency for the reservoir surface and cold water inflows to warm up. The CSHE and the Teq are derived and evaluated in Chapter 11. The Rdecay is set to zero so that residence times throughout the lake can be evaluated from the initialized arbitrary constituent concentration.

A profile slice is taken from west to east along the main stem of the lake to show the horizontal velocity profiles, the temperature profiles and the constituent concentration profiles. A similar profile is taken from south to north along the arm for the same parameters. The surface values of horizontal velocity and temperature are selected for printout. Time series are output for temperature at the surface and near the outlet at the dam.

The simulation is run at a 200-second time step over 480 hours (20 days). Printouts are made every 240 days to have an intermediate set of results for comparison to the final results to determine how close steady-state conditions are approached halfway through the simulation. The time series are printed out every 6 hours through the simulation. For constituent averages, one constituent, the arbitrary constant, is required for the residence times to be printed out.

5.1.1   Simulation Results

The main stem longitudinal slice results for the horizontal velocity component, the temperature profiles, and the residence times are given in Table 5-3. The comparison of the velocity and temperature profiles show the cold water entering with the main stem inflow and ducking under the surface to keep the bottom water cool. There is some mixing of warmer surface water down into the inflow. The velocity profiles show a surface return current from the dam back toward the inflow, and a bottom return current in a small length of the main channel. The flow toward the dam is between these two layers. Surface heating helps to produce the temperature profiles, which at the dam show a difference of 3° C between the surface and the lower layers.

A west to east longitudinal slice of velocity vectors and isotherms is shown in Figure 5-2. The figure shows that the main stem inflow mixes with the warmer surface water just as it enters the lake, but it is still cool enough that it flows toward the dam as a strong thin interflow about 2 m thick. It is located 3 m to 4 m below the surface. The temperature profile thermocline is located in the region of the interflow. As will be seen when this case is used for application with the DOD and WQDPM models, the flow pattern through the reservoir strongly influences the distribution of water quality throughout it.

The residence times in Table 5-3 are for individual cell locations within the reservoir and do not represent flushing times out of it. The local residence times are 7 days to 28 days near the main stem inflow. Proceeding down the lake to I = 6, about 2.4 km from the inflow, the residence times are larger on the surface than on the bottom. The local residence times get quite large in the deeper bottom waters near the dam.

Comparison of the 240 hour and 480 hour spatial output results, as well as the time series results, indicate that steady-state conditions are reached in 20 days. The residence times, however, suggest that a longer simulation should be made to see if they change substantially. The complete set of output results are given in the Applications subfolder for this project.

5.1.2   Effects of Groundwater Inflow

The groundwater inflow into lakes and reservoirs can have an influence on the temperature profiles and on the circulation. In this section, the project Res_TSC_01 is extended to include a groundwater inflow to show its influence. The lake project with a groundwater inflow is Res_TSC_01a.

Table 5-4 contains the portions of the Res_TSC_01 input file revised to include a groundwater inflow. The groundwater inflow is $10 \text{ m}^3/\text{s}$ into the bottom of cells with K0(I,J) extending from kgrndU = 7 downward to the bottom. The groundwater inflow is added to the outflow (Item $4) to ensure that the sum of the inflows equals the sum of the outflows as required for the reservoir analysis.

The results of including a groundwater inflow into the lake are given in Table 5-5. These results show that the groundwater inflow modifies the temperature profiles by reducing the hypolimnetic temperatures and increases the bottom circulation near the dam. The local residence times are also reduced.

5.1.3   Suggested Study Examples for Lake Project Res_TSC_01

Following are some suggested study examples.

Res_TSC_01b. Turn off the inflow and outflow by setting them to zero, and initialize the lake to a temperature of 20° C to investigate how the lake stratification develops due to surface heat exchange alone.

Res_TSC_01c. Place a surface wind on project Res_TSC_01b to investigate the effects of wind on developing the stratification and the kind of circulation that can be induced under stratified conditions.

Res_TSC_01d. Turn off surface heat exchange in Res_01b by setting the CSHE to zero. Place a surface wind on the reservoir to investigate the kind of circulation that can develop due to wind alone with no stratification.

Res_TSC_01e. Starting with Res_TSC_01, set the inflow temperatures at values higher than those found from the base simulation to determine the kind of circulation that can develop with a warm surface over flow.

Res_TSC_01f. Continue with the groundwater inflow project, Res_TSC_01a, to determine what would happen if the main stem inflow of $5 \text{ m}^3/\text{s}$ is turned off and the groundwater inflow is set to this value.

Res_TSC_01g. Beginning with the groundwater inflow project, Res_TSC_01a, determine the effects of different groundwater inflow temperatures. Assume a case where the groundwater inflow is due to warm springs at temperatures near 25° C.

5.2 Closed Loop Cooling Reservoir with Skimmer Walls

Small reservoirs are sometimes used to dissipate the waste heat from steam electric power plants. They are usually closed loop operations where the plant intake is on one part of the reservoir and the discharge is on the other. The flow from the discharge to the intake recirculates and the surface temperature builds up to a level where the dissipation of heat at the surface equals the heat rejected to the reservoir by the discharge.

There are a number of different problems to be examined for the cooling reservoir operations. One is to study the intake temperatures resulting from different kinds of operations and plant sizes to determine if they will exceed values where plant efficiency is affected. Another is to study the temperature distributions throughout the reservoir for assessment of effects on aquatic life.

5.2.1   Input Data for Cooling Reservoir Project Res_TSC_02

The cooling reservoir project is designated Res_TSC_02. The input data for this project is given in Table 5-6. There is one inflow, the plant discharge located on the southern shore approximately 1.6 km from the upper end of the reservoir. The intake is located at the southern end of the southern embayment drawing from I = 9, J = 2, and K = 2. The relationship between a steam electric plant generating capacity, heat rejection, condenser temperature rise, and pumping rate is presented later in Chapter 11, Section 11.6.

The cooling reservoir is initialized to 30° C throughout. The CSHE is set to 35 W/m^2/°C in anticipation of much higher surface temperatures. The Teq is set to 30° C so that heat dissipation at the water surface takes place only when the surface temperature exceeds the initial temperature.

Profile outputs are specified along the main stem of the reservoir and from south to north through the plant intake. Surface horizontal velocities and temperatures are specified. Time series are specified at the plant intake, near the plant discharge, and in the deeper water near the dam.

A skimmer wall is placed across the intake embayment to keep warmer surface water out and to let cooler bottom water into the intake. It extends from I = 8 at the western end to I = 11 at the eastern end along J = 5. The skimmer wall extends from Ktop = 2 at the water surface to Kbottom = 4. For the latter, there will be a maximum of 2 m of flow under it at the deepest part of the cross-section behind the skimmer wall.

5.2.2   Res_TSC_02 Simulation Results

The Res_TSC_02 detailed simulation results are given in the example applications folder. The profile of temperatures and velocity vectors along the main stem are shown in Figure 5-3. It shows high temperatures off the discharge location and decreasing surface temperatures down the reservoir. There is much cooler water on the bottom of the main stem. The velocities show a flow down the surface, a sinking near the dam, and an intermediate return flow toward the head of the reservoir. The latter has a temperature similar to that of the surface water at the dam. The return flow is a density-induced baroclinic flow.

The profile slice through the southern arm and the skimmer wall is shown in Figure 5-4. It shows relatively higher velocities under the skimmer wall into the intake embayment. There is a difference in temperature between the surface water outside the skimmer wall and at the intake of about $3.5°$ C.

The time series show that steady-state conditions are approached over the 20 days of simulation, but temperatures are still increasing slightly.

5.2.3   Cooling Reservoir Excess Temperature Computations

When analyzing cooling water discharges, excess temperature can be used as a parameter. The excess temperature is the increase in temperature due to the cooling water discharge alone. As shown in the input given in Table 5-7 for project Res_TSC_03, the computation does not require meteorologic data to evaluate the equilibrium temperature of surface heat exchange or to specify the initial temperature of the reservoir.

The resulting distribution of excess temperature and the velocity vectors are shown for the profile along the main stem in Figure 5-5. It is similar to the distribution of temperature. The excess temperature is a useful parameter for rapid assessment of the sensitivity of the results to plant pumping rates and temperature rise (heat rejection rate) as well as surface heat exchange, requiring only the specification of the CSHE to represent surface heat exchange.

5.2.4   Suggested Example Problems

Following are suggested example problems.

Res_TSC_02a or Res_TSC_03a. Perform a simulation without the skimmer wall to determine how it contributes to lowering intake temperatures.

Res_TSC_02b or Res_TSC_03b. Perform a simulation including a groundwater inflow to determine its effects on temperatures within the reservoir, on the intake temperatures, and on the temperatures at the reservoir downstream outlet. Examine the effects of outlet elevation on the temperatures.

Res_TSC_03c. The results of the simulations with the skimmer wall indicate that there is a large volume of cooler water in the bottom main channel region of the reservoir that cannot reach the intake because of the present reservoir bathymetry. Set up a new reservoir bathymetry from a copy of the existing one, name it Reservoir_01 for example, and place deeper channels extending out from under the skimmer wall to the deeper water in the channel. Perform simulations to determine the effectiveness of the channelization.

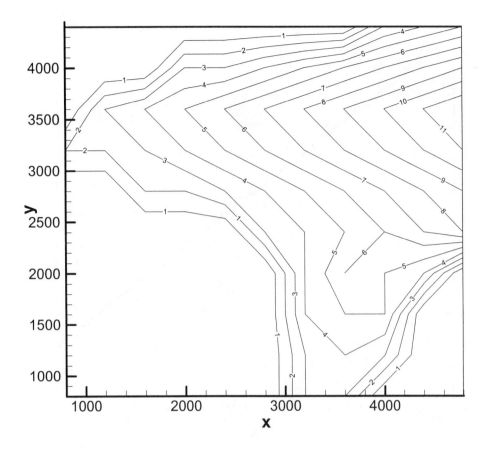

Figure 5-1. Reservoir bathymetry from Reservoir_BATH_PLT file.

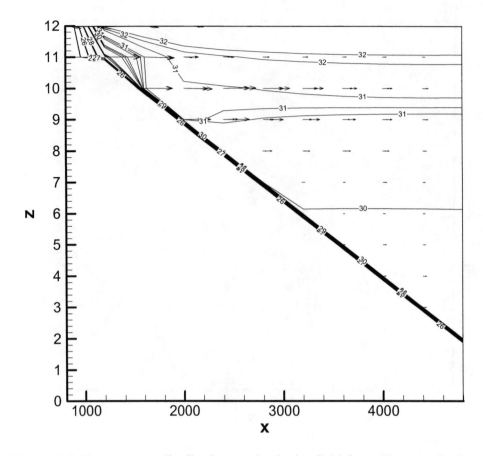

Figure 5-2. Temperature distributions and velocity field for cold water discharge into the lake.

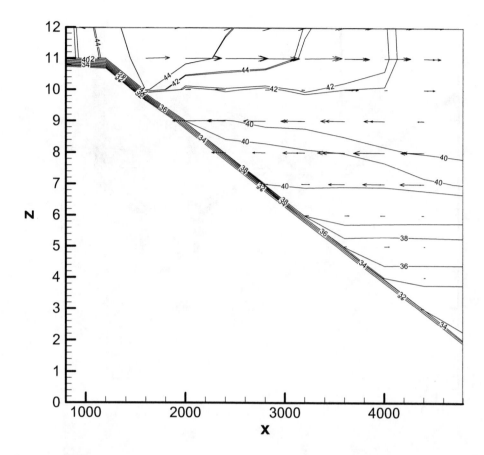

Figure 5-3. Cooling reservoir main stem temperatures and velocities for Res_TSC_02.

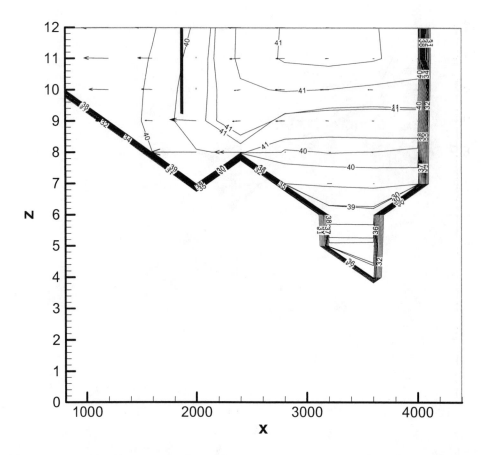

Figure 5-4. Cooling reservoir project Res_TSC_02. Profile of temperatures and velocities through intake and skimmer wall.

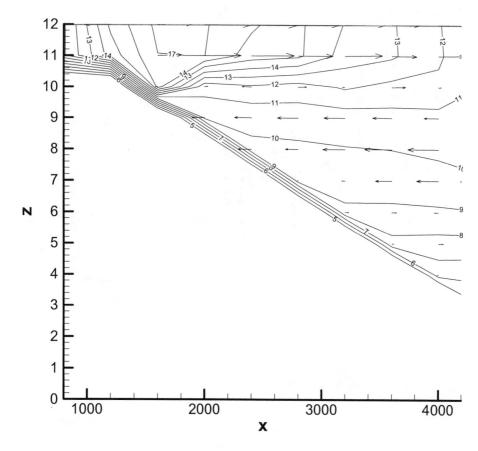

Figure 5-5. Cooling reservoir project Res_TSC_03 main stem reservoir profile of excess temperature and velocity vector distributions.

Table 5-1. Reservoir bathymetry data file.

```
Reservoir
          13      IM
          12      JM
          14      KM
     400.0000     DX
     400.0000     DY
     1.000000     DZ
           2    3    4    5    6    7    8    9    10    11    12
   11    0.0  0.0  0.0  0.0  0.0  0.0  0.0  0.0   3.0   4.0   5.0
   10    0.0  0.0  0.0  3.0  3.0  4.0  5.0  6.0   7.0   8.0   9.0
    9    0.0  3.0  4.0  5.0  6.0  7.0  8.0  9.0  10.0  11.0  12.0
    8    2.0  2.0  3.0  4.0  5.0  6.0  7.0  8.0   9.0  10.0  11.0
    7    0.0  0.0  2.0  2.0  3.0  4.0  5.0  6.0   7.0   8.0   9.0
    6    0.0  0.0  0.0  0.0  0.0  3.0  4.0  5.0   6.0   7.0   8.0
    5    0.0  0.0  0.0  0.0  0.0  0.0  4.0  6.0   5.0   4.0   0.0
    4    0.0  0.0  0.0  0.0  0.0  0.0  4.0  5.0   5.0   0.0   0.0
    3    0.0  0.0  0.0  0.0  0.0  0.0  3.0  4.0   3.0   0.0   0.0
    2    0.0  0.0  0.0  0.0  0.0  0.0  3.0  3.0   0.0   0.0   0.0
        -999     geomdelim
```

Table 5-2. Input data table for lake project Res_TSC_01.

```
Res_TSC_01
 $1.nwqm                 1
 $2.Inflow Conditions
 $ninflows               2
 $qinflow,iinflow,jinflow,kinflow
 $intake,inintake,jintake,kintake
 $temp,saln,const,cbod,on,nh3,do(d),no3,op,po4,phyt
     5.00       2       8       2
       0       0       0       0
    20.00       0.00       0.00
     0.50       8       2       2
       0       0       0       0
    10.00       0.00       0.00
 $.4 Outflow Conditions
 $noutflows               1
 $qoutflow,ioutflow,joutflow,koutflow
     5.50       12       9       11
 $4.Elevation Boundary Conditions
 $nelevation  kts              0              2
 $iewest,ieeast,jesouth,jenorth
 zmean,zamp,tmelag,tideper
 k,temp,saln,const,cbod,on,nh3,do(d),no3,op,po4,phyt for k=2,km-2
 $5.Initialize Water Quality Profiles
 $ninitial               1
 k,temp,saln,const,cbod,on,nh3,do(d),no3,op,po4,phyt for k=2,km-2
     2      30.00       0.00      100.00
     3      30.00       0.00      100.00
     4      30.00       0.00      100.00
     5      30.00       0.00      100.00
     6      30.00       0.00      100.00
     7      30.00       0.00      100.00
     8      30.00       0.00      100.00
     9      30.00       0.00      100.00
    10      30.00       0.00      100.00
    11      30.00       0.00      100.00
    12      30.00       0.00      100.00
    13      30.00       0.00      100.00
 $6.External Parameters
 Chezy, Wx,Wy,CSHE, TEQ, Rdecay, Lat.
    35.00       0.00       0.00      25.00      35.00      0.00      28.20
```

Table 5-2 (continued)

```
$7.Output Profiles
$nprofiles              2
$ipwest,ipest,jpsouth,jpnorth
$u-vel, v-vel, w-vel
$nconstituents
$I-const(1),I-const(2), I-const(3), etc
     2    12     8      8
     1     0     0
     2
     1     3
     8     8     2     10
     0     0     0
     2
     1     3
$8.Output Surfaces
$nsurfaces   $nconstituents              1              1
$U-vel   V-vel   1.0000000E+00   0.0000000E+00
$I-const(1), I-const(2), I-const(3), etc.
     1
$9.Output Time Series
$ntimser                2
$nconst, iconst, jconst,kconst
     1    12     9      2
     1    12     9     11
$10.Simulation time conditions
$dtm   tmend   200.0000         480.0
$tmeout   tmeserout   240.0      6.0
$11.Internal Boundary Locations
$nintbnd                0
$ibwest,ibeast,jbsouth,jbnorth,ktop,kbottom
$12. Constituent Averages
$nconarv                1
$nconstarvs
     3
$13. Groundwater Inflow
$ngrndwtr               0
```

Table 5-3. Lake project Res_TSC_01 results for west to east slice along the main stem of velocity profiles, temperatures profiles, and residence time in days in each model cell.

West-East U-velocity profiles, cm/sec at j=

2.	3.	4.	5.	6.	7.	8.	9.	10.	11.	12.
-0.16	-1.44	-1.99	-1.52	-1.23	-0.72	-0.48	-0.32	-0.20	-0.09	0.00
1.41	1.60	1.32	0.38	0.28	0.13	0.17	0.14	0.14	0.07	0.00
0.00	0.00	1.24	1.26	1.42	1.16	0.90	0.62	0.41	0.26	0.00
0.00	0.00	0.00	1.19	1.14	0.87	0.74	0.57	0.41	0.24	0.00
0.00	0.00	0.00	0.00	0.08	0.38	0.30	0.34	0.28	0.17	0.00
0.00	0.00	0.00	0.00	0.00	-0.18	-0.08	0.11	0.10	0.11	0.00
0.00	0.00	0.00	0.00	0.00	0.00	-0.13	0.02	0.14	0.09	0.00
0.00	0.00	0.00	0.00	0.00	0.00	0.00	0.04	0.13	0.16	0.00
0.00	0.00	0.00	0.00	0.00	0.00	0.00	0.00	0.00	0.14	0.00
0.00	0.00	0.00	0.00	0.00	0.00	0.00	0.00	0.00	0.00	0.00
0.00	0.00	0.00	0.00	0.00	0.00	0.00	0.00	0.00	0.00	0.00
0.00	0.00	0.00	0.00	0.00	0.00	0.00	0.00	0.00	0.00	0.00
0.00	0.00	0.00	0.00	0.00	0.00	0.00	0.00	0.00	0.00	0.00
0.00	0.00	0.00	0.00	0.00	0.00	0.00	0.00	0.00	0.00	0.00

West-East Constituent Profiles for temp at j=

2.	3.	4.	5.	6.	7.	8.	9.	10.	11.	12.
24.22	32.55	33.10	33.30	33.51	33.49	33.63	33.70	33.71	33.71	33.68
24.24	28.85	31.53	32.34	32.53	32.62	32.66	32.67	32.68	32.67	32.66
0.00	0.00	31.53	31.90	31.96	32.02	32.06	32.09	32.11	32.12	32.12
0.00	0.00	0.00	31.90	31.79	31.76	31.74	31.73	31.73	31.73	31.73
0.00	0.00	0.00	0.00	31.54	31.49	31.48	31.46	31.45	31.45	31.46
0.00	0.00	0.00	0.00	0.00	31.23	31.20	31.20	31.20	31.20	31.20
0.00	0.00	0.00	0.00	0.00	0.00	30.96	30.96	30.96	30.96	30.97
0.00	0.00	0.00	0.00	0.00	0.00	0.00	30.76	30.76	30.77	30.77
0.00	0.00	0.00	0.00	0.00	0.00	0.00	0.00	30.62	30.62	30.62
0.00	0.00	0.00	0.00	0.00	0.00	0.00	0.00	0.00	30.50	30.49
0.00	0.00	0.00	0.00	0.00	0.00	0.00	0.00	0.00	0.00	30.34
0.00	0.00	0.00	0.00	0.00	0.00	0.00	0.00	0.00	0.00	0.00

West-East Constituent Profiles for arvDays at j=

2.	3.	4.	5.	6.	7.	8.	9.	10.	11.	12.
7.02	34.67	42.83	48.16	54.00	56.94	63.35	70.48	79.89	88.42	96.89
7.05	16.67	28.90	36.34	41.69	45.97	50.22	55.27	63.04	72.97	84.26
0.00	0.00	28.92	32.52	34.69	37.01	40.41	44.90	51.53	61.79	74.53
0.00	0.00	0.00	32.53	34.56	36.96	41.46	46.72	53.53	63.85	77.44
0.00	0.00	0.00	0.00	39.33	45.18	50.72	59.09	66.80	77.53	90.91
0.00	0.00	0.00	0.00	0.00	62.28	69.79	81.17	91.57	102.47	115.08
0.00	0.00	0.00	0.00	0.00	0.00	100.52	112.96	127.84	140.34	153.01
0.00	0.00	0.00	0.00	0.00	0.00	0.00	159.07	175.19	190.47	205.96
0.00	0.00	0.00	0.00	0.00	0.00	0.00	0.00	234.84	250.73	269.47
0.00	0.00	0.00	0.00	0.00	0.00	0.00	0.00	0.00	336.32	355.51
0.00	0.00	0.00	0.00	0.00	0.00	0.00	0.00	0.00	0.00	556.77
0.00	0.00	0.00	0.00	0.00	0.00	0.00	0.00	0.00	0.00	0.00

Table 5-4. Portions of the lake project Res_TSC_01 input file modified to include a groundwater inflow for Project Res_TSC_01a.

```
Res_TSC_01a
 $1.nwqm              1
 $2.Inflow Conditions
 $ninflows          2
 $qinflow,iinflow,jinflow,kinflow
 $intake,inintake,jintake,kintake
 $temp,saln,const,cbod,on,nh3,do(d),no3,op,po4,phyt
      5.00    2    8    2
      0       0    0    0    0.00   0.00
     20.00    0    8    2    0.00   2
      0.50    0    8    2    0.00   2
      0       0    0    0    0.00   0
     10.00    0    0    0.00   0.00   0.00
 $.4 Outflow Conditions
 $noutflows          1
 $qoutflow,ioutflow,joutflow,koutflow
     15.50   12    9   11

(Items $5 to $12)

 $13. Groundwater Inflow
 $ngrndwtr            1
 $qgrndwtr kgrndu kgrndl
     10    8   15
  $temp   saln   const
     10.0   0.0   0.0
```

Table 5-5. Lake project with groundwater inflow Res_TSC_01a results for west to east slice along the main stem of velocity, temperature, and residence time profiles.

West-East U-velocity profiles, cm/sec at j=

2.	3.	4.	5.	6.	7.	8.	9.	10.	11.	12.
-0.15	-1.09	-1.65	-1.14	-0.82	-0.66	-0.44	-0.17	-0.08	0.04	0.00
1.40	1.57	1.41	1.03	0.65	0.65	0.45	0.34	0.29	0.07	0.00
0.00	0.00	1.33	1.54	1.26	1.11	0.90	0.62	0.41	0.29	0.00
0.00	0.00	0.00	0.41	1.45	0.91	1.01	0.75	0.51	0.27	0.00
0.00	0.00	0.00	0.00	-0.43	0.85	0.66	0.78	0.64	0.37	0.00
0.00	0.00	0.00	0.00	0.00	-1.17	-1.00	-0.24	-0.01	-0.08	0.00
0.00	0.00	0.00	0.00	0.00	0.00	-0.49	-0.46	-0.23	-0.04	0.00
0.00	0.00	0.00	0.00	0.00	0.00	0.00	0.61	0.14	0.22	0.00
0.00	0.00	0.00	0.00	0.00	0.00	0.00	0.00	0.69	0.68	0.00
0.00	0.00	0.00	0.00	0.00	0.00	0.00	0.00	0.00	0.00	0.00
0.00	0.00	0.00	0.00	0.00	0.00	0.00	0.00	0.00	0.00	0.00
0.00	0.00	0.00	0.00	0.00	0.00	0.00	0.00	0.00	0.00	0.00
0.00	0.00	0.00	0.00	0.00	0.00	0.00	0.00	0.00	0.00	0.00
0.00	0.00	0.00	0.00	0.00	0.00	0.00	0.00	0.00	0.00	0.00

West-East Constituent Profiles for temp at j=

2.	3.	4.	5.	6.	7.	8.	9.	10.	11.	12.
24.14	32.35	32.71	32.95	33.15	33.31	33.27	33.28	33.38	33.29	33.46
24.16	28.69	31.09	31.64	31.80	31.88	31.97	32.01	32.03	32.04	32.02
0.00	0.00	31.09	30.95	31.17	31.16	31.16	31.16	31.16	31.15	31.15
0.00	0.00	0.00	30.83	30.31	30.38	30.34	30.35	30.35	30.36	30.36
0.00	0.00	0.00	0.00	29.41	29.28	29.36	29.36	29.37	29.38	29.40
0.00	0.00	0.00	0.00	0.00	28.23	28.10	27.97	27.92	27.92	27.92
0.00	0.00	0.00	0.00	0.00	0.00	25.31	25.45	25.44	25.43	25.44
0.00	0.00	0.00	0.00	0.00	0.00	0.00	23.38	23.45	23.46	23.44
0.00	0.00	0.00	0.00	0.00	0.00	0.00	0.00	21.64	21.68	21.67
0.00	0.00	0.00	0.00	0.00	0.00	0.00	0.00	0.00	19.99	20.03
0.00	0.00	0.00	0.00	0.00	0.00	0.00	0.00	0.00	0.00	15.90
0.00	0.00	0.00	0.00	0.00	0.00	0.00	0.00	0.00	0.00	0.00
0.00	0.00	0.00	0.00	0.00	0.00	0.00	0.00	0.00	0.00	0.00
0.00	0.00	0.00	0.00	0.00	0.00	0.00	0.00	0.00	0.00	0.00

West-East Constituent Profiles for arvDays at j=

2.	3.	4.	5.	6.	7.	8.	9.	10.	11.	12.
6.80	31.28	36.23	40.32	44.75	49.86	51.69	55.40	61.63	66.64	74.84
6.82	15.79	25.32	29.71	33.45	37.17	40.77	45.44	51.17	58.46	66.30
0.00	0.00	25.34	27.49	29.84	32.47	35.79	39.27	43.42	50.84	57.15
0.00	0.00	0.00	27.91	30.99	32.18	35.44	38.09	40.78	45.38	50.77
0.00	0.00	0.00	0.00	33.15	33.61	35.48	38.14	39.89	42.15	45.42
0.00	0.00	0.00	0.00	0.00	34.30	34.00	34.47	35.71	38.27	40.08
0.00	0.00	0.00	0.00	0.00	0.00	24.00	26.70	27.82	28.68	29.61
0.00	0.00	0.00	0.00	0.00	0.00	0.00	21.20	22.01	22.77	23.11
0.00	0.00	0.00	0.00	0.00	0.00	0.00	0.00	18.10	18.56	18.83
0.00	0.00	0.00	0.00	0.00	0.00	0.00	0.00	0.00	15.50	15.74
0.00	0.00	0.00	0.00	0.00	0.00	0.00	0.00	0.00	0.00	10.01
0.00	0.00	0.00	0.00	0.00	0.00	0.00	0.00	0.00	0.00	0.00
0.00	0.00	0.00	0.00	0.00	0.00	0.00	0.00	0.00	0.00	0.00
0.00	0.00	0.00	0.00	0.00	0.00	0.00	0.00	0.00	0.00	0.00

Table 5-6. Input data table for cooling reservoir project Res_TSC_02.

```
Res_TSC_02
 $1.nwqm                  1
 $2.Inflow Conditions
 $ninflows               1
 $qinflow,iinflow,jinflow,kinflow
 $intake,inintake,jintake,kintake
 $temp,saln,const,cbod,on,nh3,do(d),no3,op,po4,phyt
     107.0      4      7      2
        1       9      2      2
     10.00        0.00       0.00
 $.4 Outflow Conditions
 $noutflows               1
 $qoutflow,ioutflow,joutflow,koutflow
     107.0      9      2      2
 $4.Elevation Boundary Conditions
 $nelevation  kts            0              2
 $iewest,ieeast,jesouth,jenorth
 zmean,zamp,tmelag,tideper
 k,temp,saln,const,cbod,on,nh3,do(d),no3,op,po4,phyt for k=2,km-2
 $5.Initialize Water Quality Profiles
 $ninitial               1
 k,temp,saln,const,cbod,on,nh3,do(d),no3,op,po4,phyt for k=2,km-2
        2      30.00      0.00    100.00
        3      30.00      0.00    100.00
        4      30.00      0.00    100.00
        5      30.00      0.00    100.00
        6      30.00      0.00    100.00
        7      30.00      0.00    100.00
        8      30.00      0.00    100.00
        9      30.00      0.00    100.00
       10      30.00      0.00    100.00
       11      30.00      0.00    100.00
       12      30.00      0.00    100.00
       13      30.00      0.00    100.00
 $6.External Parameters
 Chezy, Wx,Wy,CSHE, TEQ, Rdecay, Lat.
     35.00      0.00      0.00     35.00     30.00      0.00     28.20
```

Table 5-6 (continued)

```
$7.Output Profiles
 $nprofiles              2
 $ipwest,ipest,jpsouth,jpnorth
 $u-vel, v-vel, w-vel
 $nconstituents
 $I-const(1),I-const(2), I-const(3), etc
       2     12      8      8
       1      0      0
       1
       1
       9      9      2     10
       0      1      0
       1
       1
$8.Output Surfaces
$nsurfaces  $nconstituents                 1              1
$U-vel  V-vel  1.0000000E+00  0.0000000E+00
$I-const(1), I-const(2), I-const(3), etc.
       1
$9.Output Time Series
$ntimser              3
$nconst, iconst, jconst,kconst
       1      4      7      2
       1      9      2      2
       1     12      9     10
$10.Simulation time conditions
$dtm   tmend   200.0000          480.0
$tmeout  tmeserout   240.0       6.0
$11.Internal Boundary Locations
$nintbnd               1
$ibwest,ibeast,jbsouth,jbnorth,ktop,kbottom
  8   11   5   5   2   4
$12. Constituent Averages
$nconarv               0
$nconstarvs
       0
$13. Groundwater Inflow
 $ngrndwtr          0
```

Table 5-7. Input data table for project Res_TSC_03, evaluating excess temperatures for a cooling water discharge.

```
Res_TSC_03
 $1.nwqm                 1
 $2.Inflow Conditions
 $ninflows               1
 $qinflow,iinflow,jinflow,kinflow
 $intake,inintake,jintake,kintake
 $temp,saln,const,cbod,on,nh3,do(d),no3,op,po4,phyt
    107.0      4      7      2
     1      9      2      2
    10.00      0.00      0.00
 $.4 Outflow Conditions
 $noutflows              1
 $qoutflow,ioutflow,joutflow,koutflow
    107.0      9      2      2
 $4.Elevation Boundary Conditions
 $nelevation  kts        0                2
 $iewest,ieeast,jesouth,jenorth
 zmean,zamp,tmelag,tideper
 k,temp,saln,const,cbod,on,nh3,do(d),no3,op,po4,phyt for k=2,km-2
 $5.Initialize Water Quality Profiles
 $ninitial               0
 k,temp,saln,const,cbod,on,nh3,do(d),no3,op,po4,phyt for k=2,km-2
 $6.External Parameters
 Chezy, Wx,Wy,CSHE, TEQ, Rdecay, Lat.
    35.00      0.00      0.00      35.00      0.00      0.00      28.20
 $7.Output Profiles
 $nprofiles              2
 $ipwest,ipest,jpsouth,jpnorth
 $u-vel, v-vel, w-vel
 $nconstituents
 $I-const(1),I-const(2), I-const(3), etc
     2     12      8      8
     1      0      0
     1
     1
     9      9      2     10
     0      1      0
     1
     1
 $8.Output Surfaces
 $nsurfaces  $nconstituents            1                1
 $U-vel  V-vel  1.0000000E+00   0.0000000E+00
 $I-const(1), I-const(2), I-const(3), etc.
     1
 $9.Output Time Series
 $ntimser               3
 $nconst, iconst, jconst,kconst
     1      4      7      2
     1      9      2      2
     1     12      9     10
 $10.Simulation time conditions
 $dtm   tmend  200.0000          480.0
 $tmeout  tmeserout   240.0        6.0
 $11.Internal Boundary Locations
 $nintbnd                1
 $ibwest,ibeast,jbsouth,jbnorth,ktop,kbottom
  8  11  5  5  2  4
 $12. Constituent Averages
 $nconarv                0
 $nconstarvs
     0
 $13. Groundwater Inflow
 $ngrndwtr               0
```

6. APPLICATION OF THE DISSOLVED OXYGEN DEPRESSION MODEL

The DOD model allows one to determine how much an inflow will depress the dissolved oxygen within a water body due to the discharge of organic nitrogen (ON), ammonium (NH_4), and carbonaceous biochemical oxygen demand (CBOD). The DOD model is derived and discussed in detail in Chapter 12 where the CBOD oxidizes to carbon dioxide with the uptake of dissolved oxygen. The ON mineralizes to NH_4 and the NH_4 oxidizes to nitrate (NO_3) with the uptake of dissolved oxygen.

The DOD model does not require one to perform a complete dissolved oxygen study for a water body. For this reason, the DOD model should be run before the WQDPM model to get some indication about where there might be dissolved oxygen problems in the water body relative to inflows. The WQDPM model can be run for nearly the same conditions as the DOD model for a comparison of results of the former to test its setup. The reaction rates in the DOD model are temperature dependent, requiring that it always include temperature computations.

The DOD model has a nwqm = 2. Its water quality parameter table is discussed in Chapter 3, Section 3.3.2. The default parameter data file name is DOD_WQM.dat.

A number of different modeling techniques are used in this chapter to study the properties of the DOD model. The DOD model is first set up as a tank test that allows one to examine how the concentrations of each constituent change over time from a specified set of initial concentrations. It is then run for a simple stream that gives results of the longitudinal variations of the different constituents in response to a stream and point discharge inflow that can be compared with the analytical solutions for DOD given in Chapter 12. Finally, it is set up and run for the reservoir case to show the combined effects of the interaction of a complex flow field and stratification on the dissolved oxygen. These techniques apply to any water quality modeling problem.

6.1 Application to a Tank Test

In a tank test, the model geometry is set up with only four cells, each at least two layers thick. The water quality model constituent concentrations are initialized and the water quality reactions take place over time at the specified rate parameters. A surface wind is placed on the tank to keep it completely mixed. The main output of the tank test is the time series of constituent concentrations that show how the constituents are interacting and how long it takes to reach a final steady-state set of concentrations for the given rate parameters.

Another important use of the tank test is to check that the integration time step (dtm) is not too large for the chosen reaction rates. The required limit is that $Kr\Delta t \ll 1$ for a single reaction rate, Kr. There are interactions between the reactions that make the limit a complex function of almost all of the reaction rates. The tank test allows one to perform a simulation to determine if the resulting time series of the change in constituent concentrations over time are reasonable or if they change when Δt is changed. Eventually, a Δt is found below which the time series results do not change.

The bathymetric file for the tank test is given in Table 6-1. It shows four cells of depth 1 m that are divided into two layers of 0.5 m thickness each.

6.1.1   Tank Test Input File

The input file for the tank test of the DOD model is given in Table 6-2. The only items specified are the nwqm of the water quality model (2), the initial concentration of each of the DOD constituents, bottom friction and surface wind, the time step, and output times. The time series of each constituent at the center of the tank is selected for output.

The DOD default water quality rate parameters are placed in the skeleton water quality model file, Tank_DOD_01_WQM.dat, from the default parameter file DOD_WQM.dat. The DOD parameter file includes the surface wind speed from which the surface reaeration rate is computed.

6.1.2   Tank Test Results

The time series for the tank test of the DOD model is given in Figure 6-1. It shows the biochemical oxygen demand (BOD) and the ON decaying over time. It also shows the NH_4 increasing over time as the ON converts to NH_4. After the ON concentration decreases sufficiently, the NH_4 starts to decrease to NO_3. The NO_3 increases as the NH_4 is converted to it.

The dissolved oxygen depression increases from zero to a maximum as more oxygen is taken up by the decay of BOD and NH_4 than is supplied by surface reaeration. The DOD then decreases as the surface reaeration begins to dominate.

The nitrogen constituents are not exchanged at the water surface or tank bottom. The sum of the nitrogen constituents, shown in Figure 6-1, remains constant after initial startup. This demonstrates that nitrogen is conserved in the tank test.

6.1.3   Suggested Study Examples

The following suggested examples are chosen not only to illustrate properties of the tank test, but also to demonstrate properties of the DOD model.

Tank_DOD_01a. Demonstrate that the tank test conserves mass by initializing the arbitrary constituent concentration with Rdecay = 0.

Tank_DOD_01b. Determine if the tank test is sensitive to the surface area that could increase the total amount of oxygen entering it through surface reaeration.

Tank_DOD_01c. Determine the sensitivity of the tank test results to the surface reaeration wind speed specified in the tank test _DOD_WQM.dat file.

Tank_DOD_01d,e,f. Run a series of tank tests to determine if the dissolved oxygen depression is more sensitive to the initial value of BOD, ON, or NH_3. Test for the same initial concentrations of each.

Tank_DOD_01g,h,i. Run a series of sensitivity tests to the kinetic rate coefficients for the decay of BOD, the decay of ON, and the decay of NH_3.

Tank_DOD_01j. Run a series of simulations at increasing maximum time steps, dtm, to find where the results get unrealistic.

General. Nitrogen should be conserved in all the suggested study examples as shown for the project example application Tank_DOD_01. Make the same check for the rest of the projects in this chapter.

6.2 Application to a Simple Stream

Interactive properties of the hydrodynamic and water quality model can be studied initially by running a project as a simple stream. The DOD model has an analytical solution for the simple stream case that can be used to validate the numerical solutions. The technique of "folding" a simple stream on a model grid can be used to represent long tributary inflows to a larger water body while conserving computational grid space. Stream hydraulics are only approximated by this technique, limiting its use as far as providing accurate information on surface profile backwaters.

The bathymetric file for a simple stream is given in Table 6-3. It shows the length of the stream folded back and forth across the table to represent a stream running from I = 2, J = 12,11 to I = 2, J = 3,2. The stream is made two cells wide so that an elevation boundary condition can be used at the downstream end. The stream is 40 km long, 200 m wide, and 2 m deep. It is divided into a minimum of two layers.

6.2.1   Stream Input File

The stream input file is given in Table 6-4. The stream project is named str_DOD_01. It is set up for a total river inflow of 30 m^3/s, distributed as four separate inflows of 7.5 m^3/s at I = 2 for J = 11,12 and K = 2,3. The river inflow temperature is 30° C, but it carries no other constituents into the system. Similarly, a total facility discharge of 15 m^3/s enters at I = 3, distributed as four separate inflows of 3.75 m^3/s each for J = 11,12 and K = 2,3. The inflows are distributed uniformly over the cross-section of the river to ensure complete mixing between the river and the facility discharge. The facility discharges at the indicated temperature and concentrations for each of the constituents in the DOD model.

The downstream end of the stream reach is controlled by an elevation boundary condition. The temperature profile at the elevation boundary is initialized in case there is any backflow during initial model spin up. The temperature in the stream is initialized to 30° C to eliminate the necessity to spin up the temperatures along it.

Profile slices are taken for the first reach and last reach of the stream. The program will print out the surface elevation along the reach, the velocity in each layer, and the DOD in each layer. These can be checked to see how much vertical shear is developed and to see if the DOD is truly vertically mixed.

The surface distribution specifies the printout of all the DOD model constituents at the end of the simulation. Time series in the last cell before the two elevation boundary cells are specified for each constituent and surface velocity. The time series of surface elevation is specified at the head of the stream. These are chosen as checks on the approach to steady-state conditions.

6.2.2   Stream Water Quality Parameter File

The default water quality parameters from DOD_WQM.dat should be placed in skeleton water quality parameter file Str_DOD_01_WQM.dat.

The water quality parameter file for the project requires a reaeration wind speed, Wad. The Mackay (1980) formula is used to evaluate surface reaeration as a function of wind speed. It applies to relatively deep water bodies with large surface areas. Stream reaeration coefficients depend on stream velocity and depth as in the Churchill et al. (1962) formula. Equating the two, as shown in Chapter 12, Section 12.7, gives an *effective* reaeration wind speed that can be used on streams of

$$Wad = 19.7 * U^{0.64} / H^{0.44} \qquad\qquad (6.1)$$

where U is the average stream velocity computed by dividing the river flow by its cross-section and H is the average stream depth.

The total stream flow downstream of the facility is 45 m^3/s and the bathymetry gives a cross-section of 400 m^2. These combine to give a mean stream velocity of 0.113 m/s at a depth of 2 m. The effective wind speed, from Equation 6.1, is 3.6 m/s.

6.2.3   Stream Results

The profile of velocity and DOD and the longitudinal distribution of DOD constituents are given in Table 6-5. The profiles show that the horizontal velocities are different between the two layers and there is velocity shear. The DOD is uniform between the two layers due to sufficient velocity shear.

The average of the top and bottom velocities at each location downstream of the facility discharge varies around the mean cross-sectional velocity of 0.113 m/s. The top and bottom averages are slightly less immediately below the discharge because of the increased surface elevation, and approach it near the downstream end.

Along the stream, the CBOD and NO concentrations decrease with distance. There is an increase in NH_3 due to the mineralization of NO to NH_3, followed by a decrease as the NH_3 is oxidized to NO_3. The DOD first increases due to the oxidation of CBOD and NH_3, followed by a decrease due to surface rearation becoming dominate. The NO_3 increases with distance from the discharge.

The time series results, shown in str_DOD_01_TSO.dat, show the concentrations of all constituents increasing with time and all reaching a steady-state value after approximately 185 hours of simulation. The CBOD takes the longest to reach a steady-state value. The elevation near the discharge and the velocities reach steady state quite rapidly.

6.2.4   *Suggested Study Examples*

Following are some suggested study examples.

Str_DOD_01a. Increase the facility discharge temperature to 40° C to determine the effects of a cooling water discharge on downstream DOD. Compare the computed longitudinal temperature distributions with the analytical solution to the one-dimensional temperature equation given in Chapter 11, Section 11.5.

Str_DOD_01b. Set the river inflow to zero so that the only flow in the stream is the facility discharge and so that more pronounced constituent concentration gradients are produced. Also, set the facility discharge concentration of ON to zero. Compare the computed longitudinal distributions of CBOD, NH_4, and DOD to their analytical solutions given in Chapter 12. Evaluate the time in the analytical solutions as a travel time where t = X/U and X is the longitudinal distance from the discharge and U is the mean stream velocity.

Demonstrate for all three projects, str_DOD_01, str_DOD_01a, and str_DOD_01b, that nitrogen is conserved between the facility inflow and the outflow. Do this by estimating the mass flow rate of each nitrogen constituent into the stream from the discharge as the product of the facility flow and constituent concentration, and comparing it with the mass flow rate out of the lower end of the stream.

Discuss the differences between using the tank test and the simple stream to examine the properties of a water quality model.

6.3 Application to a Reservoir

The reservoir project Res_TSC_01 given in Chapter 5 is used here to develop an example application of the DOD model to a reservoir. The reservoir project was chosen to show the interaction of the flow field and the water quality modeling results, and in particular the effects of stratification. The bathymetry of the reservoir is as given in Table 5-1.

6.3.1   Reservoir Input File

The reservoir project is named Res_DOD_01. Its input file, Res_DOD_01_INP.dat, can be found in the example application folder. It is set up only for a main stem river inflow. The inflow has relatively high concentrations of water quality constituents that might be typical of runoff from a heavily used watershed. Except for temperature, all the water quality constituents are initialized to zero.

6.3.2   Reservoir Results

The project time series results are found in the Res_DOD_01_TSO.dat file. They are printed out at the dam outlet to determine if the simulation time is long enough for steady-state conditions to be reached at that location. Reaching steady-state conditions at the dam does not necessarily mean they are reached at all locations.

The main stem profiles for velocity, temperature, and all the water quality constituents are found in the Res_DOD_01_SPO.dat file. The main constituent of interest, DOD, is given as a profile plot in Figure 6-2. It shows that as the flow proceeds through the reservoir the DOD decreases toward the dam and vertically downward toward the outlet. The watershed loadings are sufficiently large to have a significant potential impact on the reservoir dissolved oxygen.

As indicated in the project spatial output file, the other water quality constituents follow the expected change in concentrations as they pass through the reservoir.

6.3.3   Suggested Study Examples

Following are some suggested study examples.

Res_DOD_01a. Vary the inflow concentrations to find which water shed constituent to control or reduce so that the dissolved oxygen depression is less than 1.5 mg/l throughout the length of the reservoir.

Res_DOD_01b. Place an interior boundary three cells up reservoir of the dam extending from the bottom to a few layers below the water surface so that it acts as an overflow weir. Test to determine if this reduces the DOD in the reservoir.

Res_DOD_01c. Vary the height of the outflow at the dam to see how this changes the DOD in the reservoir.

Res_DOD_01d. Using the interior boundaries, place a marina near the mouth of the southern arm. Place one interior boundary extending from ibwest = 9 to ibeast = 10 along jbsouth = jbnorth = 4. Place the second along ibwest = ibeast = 8 extending from jbsouth = 3 to jbsouth = 4. Place an inflow to the southern arm of 3 m^3/s at I = 8, J = 2, and K = 2 with the same inflow constituents as in the main arm. Set up the input data file for the computation of the distribution of flushing time throughout the lake. Simulating the project with and without the marina breakwaters, determine the effect of the latter on flushing time in the southern arm, and on the water quality as indicated by the DOD.

Discuss what happens to the ON, NH_3, NO_3, and DOD if the closed loop cooling lake Res_TSC_02 is initialized and run for these constituents.

General. For all the reservoir projects, determine if nitrogen is conserved by comparing its mass inflow at the head water river with its mass outflow at the dam. How good a measure of the approach to steady-state conditions is this balance?

Est_DOD_01. Starting with Est_TSC_01, set up an estuarine case with the same inflow rate, temperature, and constituent concentrations as used in the reservoir. Turn off the facility inflow and withdrawal. Initialize temperatures within the reservoir and at the tidal boundary to 30° C, and use the same salinity profile. Initialize the water quality constituents to zero. Compare the DOD distributions in the reservoir with those in the estuary and explain the differences.

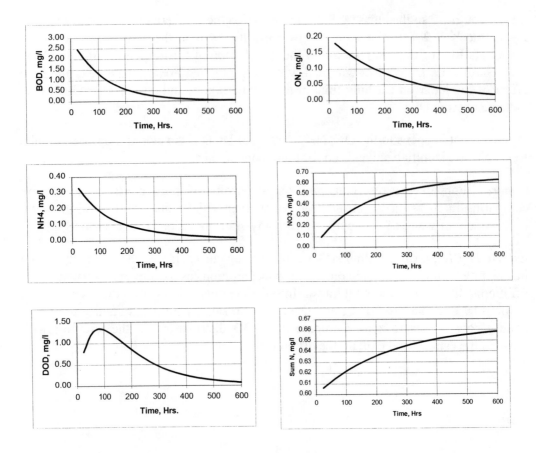

Figure 6-1. Tank test project Tank_DOD_01 showing time series of BOD, ON, NH4, NO3 and DOD and sum of N. From initialized concentrations. Notation in abbreviations.

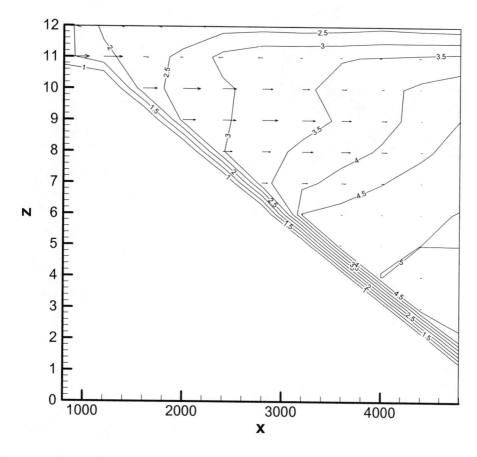

Figure 6-2. Reservoir project Res_DOD_01 main stem profile slice of dissolved oxygen depression, mg/l and velocities.

Table 6-1. Bathymetric file for tank test.

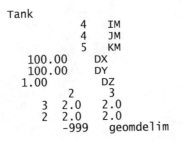

```
Tank
              4    IM
              4    JM
              5    KM
      100.00    DX
      100.00    DY
      1.00      DZ
              2    3
      3   2.0    2.0
      2   2.0    2.0
            -999    geomdelim
```

Table 6-2. Input data for tank test Project Tank_DOD_01.

```
Tank_DOD_01
 $1.nwqm              2
 $2.Inflow Conditions
 $ninflows            0
 $qinflow,iinflow,jinflow,kinflow
 $intake,inintake,jintake,kintake
 $temp,saln,const,cbod,on,nh3,do(d),no3,op,po4,phyt
 $.4 Outflow Conditions
 $noutflows           0
 $qoutflow,ioutflow,joutflow,koutflow
 $4.Elevation Boundary Conditions
 $nelevation  kts           0              2
 $iewest,ieeast,jesouth,jenorth
 zmean,zamp,tmelag,tideper
 k,temp,saln,const,cbod,on,nh3,do(d),no3,op,po4,phyt for k=2,km-2
 $5.Initialize Water Quality Profiles
 $ninitial            1
 k,temp,saln,const,cbod,on,nh3,do(d),no3,op,po4,phyt for k=2,km-2
    2     30.00       0.00       0.00       3.00       0.20       0.40       0.00
    3     30.00       0,00       0.00       3.00       0.20       0.40       0.00
 $6.External Parameters
 Chezy, Wx,Wy,CSHE, TEQ, Rdecay, Lat.
    35.00       3.00       0.00       0.00      30.00       0.00       0.00
 $7.Output Profiles
 $nprofiles           0
 $ipwest,ipest,jpsouth,jpnorth
 $u-vel, v-vel, w-vel
 $nconstituents
 $I-const(1),I-const(2), I-const(3), etc
 $8.Output Surfaces
 $nsurfaces   $nconstituents          0              0
 $U-vel   V-vel   0.0000000E+00   0.0000000E+00
 $I-const(1), I-const(2), I-const(3), etc.
 $9.Output Time Series
 $ntimser             4
 $nconst, iconst, jconst,kconst
       4      2      2      2
       5      2      2      2
       6      2      2      2
       7      2      2      2
 $10.Simulation time conditions
 $dtm    tmend    600.0000        960.0000
 $tmeout   tmeserout   960.0000       24.000000
 $11.Internal Boundary Locations
 $nintbnd             0
 $ibwest,ibeast,jbsouth,jbnorth,ktop,kbottom
 $12. Constituent Averages
 $nconarv             0
 $nconstarvs
 $13. Groundwater Inflow
 $ngrndwtr          0
```

Table 6-3. Bathymetry for a simple stream.

```
stream_01
            12    IM
            13    JM
             5    KM
     1000.0000    DX
     100.00000    DY
     1.000000     DZ
             2     3     4     5     6     7     8     9    10    11
     12    2.0   2.0   2.0   2.0   2.0   2.0   2.0   2.0   2.0   2.0
     11    2.0   2.0   2.0   2.0   2.0   2.0   2.0   2.0   2.0   2.0
     10    0.0   0.0   0.0   0.0   0.0   0.0   0.0   0.0   0.0   2.0
      9    2.0   2.0   2.0   2.0   2.0   2.0   2.0   2.0   2.0   2.0
      8    2.0   2.0   2.0   2.0   2.0   2.0   2.0   2.0   2.0   2.0
      7    2.0   0.0   0.0   0.0   0.0   0.0   0.0   0.0   0.0   0.0
      6    2.0   2.0   2.0   2.0   2.0   2.0   2.0   2.0   2.0   2.0
      5    2.0   2.0   2.0   2.0   2.0   2.0   2.0   2.0   2.0   2.0
      4    0.0   0.0   0.0   0.0   0.0   0.0   0.0   0.0   0.0   2.0
      3    2.0   2.0   2.0   2.0   2.0   2.0   2.0   2.0   2.0   2.0
      2    2.0   2.0   2.0   2.0   2.0   2.0   2.0   2.0   2.0   2.0
          -999    geomdelim
```

Table 6-4. Input file for stream project Str_DOD_01.

```
Str_DOD_01
 $1.nwqm                  2
 $2.Inflow Conditions
 $ninflows                   8
 $qinflow,iinflow,jinflow,kinflow
 $intake,inintake,jintake,kintake
 $temp,saln,const,cbod,on,nh3,do(d),no3,op,po4,phyt
    7.500      2       12       2
       0     0      0       0
    30.00     0.00       0.00     0.00     0.00     0.00     0.00   0.00
    7.500      2       11       2
       0     0      0       0
    30.00     0.00       0.00     0.00     0.00     0.00     0.00   0.00
    7.500      2       12       3
       0     0      0       0
    30.00     0.00       0.00     0.00     0.00     0.00     0.00   0.00
    7.500      2       11       3
       0     0      0       0
    30.00     0.00       0.00     0.00     0.00     0.00     0.00   0.00
    3.75       3       12       2
       0     0      0       0
    30.00     0.00       0.00     6.00     2.00     4.00     0.00   0.00
    3.75       3       11       2
       0     0      0       0
    30.00     0.00       0.00     6.00     2.00     4.00     0.00   0.00
    3.75       3       12       3
       0     0      0       0
    30.00     0.00       0.00     6.00     2.00     4.00     0.00   0.00
    3.75       3       11       3
       0     0      0       0
    30.00     0.00       0.00     6.00     2.00     4.00     0.00   0.00
 $.4 Outflow Conditions
 $noutflows                0
 $qoutflow,ioutflow,joutflow,koutflow
 $4.Elevation Boundary Conditions
 $nelevation   kts            1            2
 $iewest,ieeast,jesouth,jenorth
 zmean,zamp,tmelag,tideper
 k,temp,saln,const,cbod,on,nh3,do(d),no3,op,po4,phyt for k=2,km-2
     2      2      2       3
    0.00     0.00      0.00      0.00
     2      30.00      0.00      0.00     0.00     0.00     0.00    0.00  0.00
     3      30.00      0.00      0.00     0.00     0.00     0.00    0.00  0.00
 $5.Initialize Water Quality Profiles
 $ninitial             1
 k,temp,saln,const,cbod,on,nh3,do(d),no3,op,po4,phyt for k=2,km-2
     2      30.00      0.00   1000.00     0.00     0.00     0.00    0.00  0.00
     3      30.00      0.00   1000.00     0.00     0.00     0.00    0.00  0.00
 $6.External Parameters
 Chezy, Wx,Wy,CSHE, TEQ, Rdecay, Lat.
    17.00      0.00      0.00    25.00     30.00     0.00     35.00
```

Table 6-4 (continued)

```
$7.Output Profiles
$nprofiles              2
$ipwest,ipest,jpsouth,jpnorth
$u-vel, v-vel, w-vel
$nconstituents
$I-const(1),I-const(2), I-const(3), etc
 2  11    12  12
 1   0    0
 1
 7
 2  11    2   2
 1   0    0
 1
 7
$8.Output Surfaces
$nsurfaces   $nconstituents                 1              5
$U-vel   V-vel   0.0000000E+00   0.0000000E+00
$I-const(1), I-const(2), I-const(3), etc.
   4       5       6       7       8
$9.Output Time Series
$ntimser             8
$nconst, iconst, jconst,kconst
    4       5       2       2
    5       5       2       2
    6       5       2       2
    7       5       2       2
    8       5       2       2
   20       5      12       2
   21       8      12       2
   21       8      12       2
$10.Simulation time conditions
$dtm    tmend    120.0000            501.00000
$tmeout  tmeserout   500.00000           5.000000
$11.Internal Boundary Locations
$nintbnd              0
$ibwest,ibeast,jbsouth,jbnorth,ktop,kbottom
$12. Constituent Averages
$nconarv              0
$nconstarvs
$13. Groundwater Inflow
$ngrndwtr         0
```

Table 6-5. Stream project Str_DOD_01 profiles of velocity and dissolved oxygen deficit and longitudinal distributions of carbonaceous biochemical oxygen demand, organic nitrogen, ammonia, dissolved oxygen deficit, and nitrate.

```
West-East water surface elevations, cm at j=          12
  35.52   35.08   34.24   33.43   32.62   31.80   30.97   30.13   29.28   28.41

West-East U-velocity profiles, cm/sec at j=          12
    2.      3.      4.      5.      6.      7.      8.      9.     10.     11.
  7.27   11.21   11.27   11.32   11.37   11.42   11.47   11.52   11.57    0.00
  5.17    7.40    7.41    7.44    7.47    7.50    7.53    7.56    7.59    0.00
  0.00    0.00    0.00    0.00    0.00    0.00    0.00    0.00    0.00    0.00

West-East Constituent Profiles forDOD  at j=          12
    2.      3.      4.      5.      6.      7.      8.      9.     10.     11.
  0.01    0.22    0.42    0.60    0.77    0.93    1.08    1.22    1.34    1.46
  0.01    0.23    0.43    0.61    0.78    0.94    1.09    1.22    1.35    1.46
  0.00    0.00    0.00    0.00    0.00    0.00    0.00    0.00    0.00    0.00

West-East water surface elevations, cm at j=           2
  0.00    0.00    1.35    2.68    3.97    5.24    6.48    7.69    8.89   10.18

West-East U-velocity profiles, cm/sec at j=            2
    2.      3.      4.      5.      6.      7.      8.      9.     10.     11.
-11.57  -12.84  -13.46  -13.36  -13.25  -13.16  -13.06  -13.00  -13.25    0.00
 -7.47   -8.62   -8.76   -8.70   -8.63   -8.57   -8.51   -8.47   -8.63    0.00
  0.00    0.00    0.00    0.00    0.00    0.00    0.00    0.00    0.00    0.00

West-East Constituent Profiles forDOD  at j=           2
    2.      3.      4.      5.      6.      7.      8.      9.     10.     11.
  0.00    0.00    2.36    2.40    2.41    2.41    2.42    2.42    2.43    2.43
  0.00    0.00    2.37    2.41    2.42    2.42    2.43    2.43    2.44    2.44
  0.00    0.00    0.00    0.00    0.00    0.00    0.00    0.00    0.00    0.00

Surface Constituent for   CBOD
  0.03    1.88    1.85    1.81    1.77    1.73    1.69    1.65    1.62    1.58
  0.03    1.88    1.85    1.81    1.77    1.73    1.69    1.65    1.62    1.58
  0.00    0.00    0.00    0.00    0.00    0.00    0.00    0.00    0.00    1.56
  1.26    1.28    1.31    1.34    1.37    1.40    1.44    1.47    1.51    1.54
  1.25    1.28    1.31    1.33    1.36    1.39    1.42    1.45    1.48    1.51
  1.24    0.00    0.00    0.00    0.00    0.00    0.00    0.00    0.00    0.00
  1.22    1.20    1.17    1.14    1.12    1.09    1.07    1.05    1.02    1.00
  1.20    1.18    1.15    1.13    1.11    1.08    1.06    1.04    1.02    1.00
  0.00    0.00    0.00    0.00    0.00    0.00    0.00    0.00    0.00    0.99
  0.00    0.00    0.83    0.86    0.88    0.90    0.92    0.94    0.96    0.98
  0.00    0.00    0.83    0.85    0.87    0.89    0.91    0.92    0.94    0.96

Surface Constituent for   ON
  0.01    0.64    0.63    0.62    0.62    0.61    0.60    0.60    0.59    0.58
  0.01    0.64    0.63    0.62    0.62    0.61    0.60    0.60    0.59    0.58
  0.00    0.00    0.00    0.00    0.00    0.00    0.00    0.00    0.00    0.58
  0.52    0.52    0.53    0.54    0.54    0.55    0.56    0.56    0.57    0.58
  0.52    0.52    0.53    0.53    0.54    0.55    0.55    0.56    0.56    0.57
  0.52    0.00    0.00    0.00    0.00    0.00    0.00    0.00    0.00    0.00
  0.51    0.51    0.50    0.49    0.49    0.48    0.48    0.47    0.47    0.46
  0.51    0.50    0.50    0.49    0.49    0.48    0.48    0.47    0.47    0.46
  0.00    0.00    0.00    0.00    0.00    0.00    0.00    0.00    0.00    0.46
  0.00    0.00    0.42    0.43    0.43    0.44    0.44    0.45    0.45    0.46
  0.00    0.00    0.42    0.43    0.43    0.43    0.44    0.44    0.45    0.45
```

Table 6-5 (continued)

```
Surface Constituent for   NH3
    0.02    1.26    1.24    1.21    1.18    1.16    1.13    1.11    1.08    1.06
    0.02    1.26    1.24    1.21    1.18    1.16    1.13    1.11    1.08    1.06
    0.00    0.00    0.00    0.00    0.00    0.00    0.00    0.00    0.00    1.05
    0.85    0.87    0.89    0.91    0.92    0.95    0.97    0.99    1.01    1.03
    0.85    0.86    0.88    0.90    0.92    0.94    0.96    0.98    0.99    1.01
    0.84    0.00    0.00    0.00    0.00    0.00    0.00    0.00    0.00    0.00
    0.83    0.81    0.80    0.78    0.76    0.75    0.74    0.72    0.71    0.69
    0.82    0.80    0.79    0.77    0.76    0.74    0.73    0.72    0.70    0.69
    0.00    0.00    0.00    0.00    0.00    0.00    0.00    0.00    0.00    0.69
    0.00    0.00    0.59    0.61    0.62    0.63    0.64    0.65    0.67    0.68
    0.00    0.00    0.59    0.60    0.61    0.62    0.63    0.65    0.66    0.67

Surface Constituent for   DOD
    0.01    0.22    0.42    0.60    0.77    0.93    1.08    1.22    1.34    1.46
    0.01    0.22    0.42    0.60    0.77    0.93    1.08    1.22    1.34    1.46
    0.00    0.00    0.00    0.00    0.00    0.00    0.00    0.00    0.00    1.51
    2.21    2.17    2.12    2.06    2.00    1.93    1.86    1.77    1.68    1.57
    2.21    2.18    2.13    2.08    2.02    1.96    1.90    1.82    1.74    1.66
    2.23    0.00    0.00    0.00    0.00    0.00    0.00    0.00    0.00    0.00
    2.25    2.29    2.32    2.35    2.37    2.38    2.40    2.41    2.42    2.42
    2.29    2.31    2.34    2.36    2.38    2.39    2.40    2.41    2.42    2.42
    0.00    0.00    0.00    0.00    0.00    0.00    0.00    0.00    0.00    2.42
    0.00    0.00    2.37    2.40    2.41    2.42    2.42    2.42    2.43    2.43
    0.00    0.00    2.36    2.40    2.41    2.41    2.42    2.42    2.43    2.43

Surface Constituent for   NO3
    0.00    0.04    0.08    0.12    0.15    0.19    0.22    0.26    0.29    0.32
    0.00    0.04    0.08    0.12    0.15    0.19    0.22    0.26    0.29    0.32
    0.00    0.00    0.00    0.00    0.00    0.00    0.00    0.00    0.00    0.34
    0.61    0.59    0.56    0.53    0.51    0.48    0.45    0.42    0.39    0.35
    0.62    0.59    0.57    0.54    0.52    0.49    0.46    0.44    0.41    0.38
    0.63    0.00    0.00    0.00    0.00    0.00    0.00    0.00    0.00    0.00
    0.64    0.67    0.69    0.72    0.74    0.76    0.78    0.80    0.82    0.84
    0.66    0.68    0.71    0.73    0.75    0.77    0.79    0.81    0.83    0.85
    0.00    0.00    0.00    0.00    0.00    0.00    0.00    0.00    0.00    0.86
    0.00    0.00    0.98    0.98    0.96    0.94    0.92    0.91    0.89    0.87
    0.00    0.00    0.99    0.98    0.97    0.95    0.93    0.92    0.90    0.88
```

CBOD, carbonaceous biochemical oxygen demand; DOD, dissolved oxygen deficit; NH_3, ammonia; NO_3, nitrate; ON, organic nitrogen.

7. APPLICATION OF THE WATER QUALITY DISSOLVED PARTICULATE EUTROPHICATION MODEL

As illustrated in Figure 13-1, the WQDPM eutrophication model cycles nutrients between phytoplankton and dissolved and particulate forms of carbon, nitrogen, and phosphorous. Zooplankton grazing performs part of the cycling. Dissolved oxygen is taken up at different stages in the reduction of the carbon, nitrogen, and phosphorous to simpler forms and is re-supplied by photosynthesis and surface reaeration. The particulate constituents that settle to the bottom contribute to the benthic oxygen demand and provide the material for nutrients that are released from the sediment back into the water column. The processes and nutrient cycling that take place in WQDPM are first examined in this chapter using a series of tank tests. The tank tests show that WQDPM can be set up as a very simple water quality model and extended through to the complete, relatively complex model.

In addition to the constituent concentrations through time, another parameter obtained throughout the WQDPM simulations is the net phytoplankton growth rate computed as

$$Algrt = [(1\text{-}fe)gp - gr - dr - dd - (Kgmicro+Kgmacro)] \qquad (7.1)$$

where:

fe = the excretion fraction of phytoplankton as defined in Table 13-3

gp = the growth rate after the maximum growth rate (k1c) is modified by nutrient and light limitations and for temperature, given in Table 13-2, Equation 13.4.2

dr = the respiration rate after the maximum respiration rate (k1r) is adjusted for temperature, given in Table 13-2, Equation 13.4.4

dd = the death rate and is the same as the maximum death rate (k1d), given in Table 13-2, Equation 13.4.5

Kgmicro = the microzooplankton grazing rate, as defined in Table 13-3

Kgmacro= macrozooplankton grazing rate, as defined in Table 13-3

These terms are included in the phytoplankton rate relationship given in Table 13-2, Equation 13.4.1. Algrt provides an indication of when algae are growing and dying in the tank tests, and where algae are growing and dying throughout a water body.

Along with the tank tests, in this chapter WQDPM will be applied to the flow through a lake as well as a recirculating cooling water reservoir. The first illustrates the use of preliminary modeling to examine different methods of initializing the water quality distributions at the beginning of the simulation. The second illustrates the use of preliminary modeling to examine problems that arise in specifying discharge conditions. Both provide examples of the kinds of water quality distributions that can be expected in lakes and reservoirs.

7.1 Tank Tests of Water Quality Dissolved Particulate Model Properties

A series of five tank tests are set up in this section to illustrate the increasing stages of complexity of the WQDPM model. The WQDPM parameters applied to each test are given in Table 7-1. The tank tests range from the conditions used in the DOD tests through to simulations with algae but without settling, and then with all processes acting.

The tank tests are set up for the input data given in Table 7-2. The WQDPM model number is nwqm = 3. The number of constituents included for inflows, elevation boundary conditions, and initialization extend beyond the DOD model to include organic phosphorous (OP), phosphate (PO_4), phytoplankton (Phyt), particulate carbonaceous biochemical oxygen demand (CBOD_P), particulate organic nitrogen (ON_P), and particulate organic phosporus (OP_P) . The constituents are initialized to the same concentrations as used in the DOD tests through NO_3. The OP and PO_4 are added to the tanks so that the algal growth limitation due to decreasing PO_4 levels is not reached too early in the simulation for cases 01d and 01e in Table 7-1. A surface wind of 3 m/s is placed on the tank to keep it fully mixed. The results of the five cases are given as subcases in the example application folder Tank_WQDPM_01.

7.1.1   Comparison to the Dissolved Oxygen Deficit Model Results

The project Tank_WQDPM_01a is set up here to duplicate the results of the DOD tank test. The project Tank_WQDPM_01b gives similar results for dissolved oxygen even though it includes the dissolved organic phosporus (OP_D) and PO_4 relationships. The latter is used for comparison so that the conservation of nitrogen and of phosphorous can be checked out.

The results of the Tank_WQDPM_01a are given in Figure 7-1. The results are almost identical to those shown for the DOD tank test in Figure 6-1. The dissolved oxygen is depressed about 1.5 mg/l below saturation and corresponds to the maximum DOD. Both the nitrogen and phosphorous are conserved. The comparison demonstrates that the WQDPM model reduces to an elementary dissolved oxygen model when no particulate components or phytoplankton are included.

7.1.2   Phytoplankton with Dissolved Constituents Only

The project Tank_WQDPM_01c includes all dissolved constituents along with phytoplankton. This setup is similar to many early eutrophication models. It is initialized for the constituent concentrations in Table 7-2.

The results are shown in Figure 7-2. They demonstrate that the phytoplankton have a big influence on increasing the dissolved biochemical oxygen demand (BOD)_d and ON_d after initial decay. The NH_4 decays to NO_3 and is taken up by the phytoplankton. The NO_3 increases with time until after the NH_4 reaches a low concentration, then is taken up by the phytoplankton. The changes in NH_4 and NO_3

show the preference by phytoplankton for the former until it is exhausted. The nitrogen in solution is not conserved because a portion of it is stored in the phytoplankton.

The dissolved organic phosporus (OP)_d decays to PO_4 with a slight increase due to release by the phytoplankton. The PO_4 increases to a steady concentration, and then is decreased by algal uptake. The phosphorous in solution is not conserved because a portion of it is stored in the phytoplankton.

The phytoplankton begins growing at approximately 300 hours, and reaches a peak density near 400 hours. It is during the approach to and at the time of the peak density that the major changes take place in the different constituent concentrations. The change in maximum algal densities occurs when the growth rate changes. The sudden decrease in growth rate is due to the decrease of NH_4 and NO_3 to near zero, which as shown by Equation 13.4.2 in Table 13-2 decreases the nitrogen limit on growth to zero.

7.1.3 *Phytoplankton with all Constituents, Settling, and Sediment Exchange*

Use of all of the components in WQDPM is illustrated in this section for the project Tank_WQDPM_01e. The Pamatmat (1971) sediment exchange relationship is used for this application. The WQDPM parameters for the project are given in Table 7-1 and the input data are given in Table 7-2. The concentrations of the particulate constituents are initialized at zero and are produced by zooplankton grazing, and death and excretion from phytoplankton.

The results of the simulation are shown in Figure 7-3. Most of the dissolved constituents behave similarly to those for Tank_WQDPM_01c described in Section 7.1.2. Particulate components eventually develop and settle out. The effect of the bottom release from the settled constituents is to increase the total nitrogen and phosphorous to higher values toward the end of the simulation than for the tank test with only dissolved constituents.

The dissolved oxygen concentration with particulates and bottom exchange has a slightly lower maximum value occurring later in the simulation than the project with dissolved constituents only.

7.1.4 *Example Study Problems*

For the example study problems, it is suggested that a new series named Tank_wqdpm_02 be started. The example study problems also include simple stream cases whose setup follows the example application project Str_DOD_01.

Tank_wqdpm_02a. Beginning with project Tank_wqdpm_01, set up a case to test how well the WQDPM produces the saturation value of dissolved oxygen for the specified initial temperature of the tank. Remember to set all decay rates to zero so that no dissolved oxygen is lost through these processes.

Tank_wqdpm_02b. Perform sensitivity tests to initialize the tank at higher levels of OP_d and PO$_4$ to determine if there is any change in the results from the different Tank_wqdpm_01 cases.

Tank_wqdpm_02c. Try different values of the phytoplankton growth, death, respiration, and settling rates within the indicated ranges of accepted values to determine how each affects the water quality modeling.

Str_wqdpm_01a. Set up Case 01 in Table 7-1 for application of a dissolved constituent model to a stream using the same input flow rates and constituent concentrations as given in Str_DOD_01. Initialize the dissolved oxygen to 8 mg/l. Expand Equation 12.9 to a dissolved oxygen equation by letting D = Cs-C, where Cs is the saturation dissolved oxygen concentration at the stream temperature and C is the dissolved oxygen concentration. Compare the numerical and analytical solution.

Str_wqdpm_01b. For Case 01 in Table 7-1 and using the same setup as in Str_wqdpm_01a, try different values of measured sediment oxygen demand (SODm) to determine its effect on the dissolved oxygen distribution along the stream. Is there an analytical solution for the simple stream case for constant SODm?

7.2 Reservoir Applications of the Water Quality Dissolved Particulate Model

In this section, the WQDPM model is applied to two reservoir projects. The first is the recreational lake with inflows and outflows. The second is the closed loop cooling reservoir. Each presents different problems with initialization of the water quality constituents and the inflows. The reservoir bathymetry is shown in Figure 5-1.

7.2.1   Lake with Inflows and Outflows

The input data file for the lake with inflows and outflows is given in Table 7-3. The headwater inflow to the lake is fairly heavily loaded with dissolved CBOD inorganic phosphorous (PO$_4$). The lake is initialized only for temperature and dissolved oxygen. The remaining constituents become established with the inflow. A 100 ug/l virtual dye concentration is maintained in the inflow to trace its progress through the lake. The Teq is set at 35° C, so that the lake warms up and stratification becomes established over the period of simulation. The inflow to the lake and the meteorologic conditions are given as mean monthly values.

The velocity and water quality distribution is examined on a longitudinal–vertical slice down the centerline of the lake. Time series are taken for all constituents on the surface near the dam to determine the approach to steady-state conditions. The simulation time is for 720 hours or approximately 1 month, and the time series are output every 12 hours.

The results of the simulation are given in the example applications folder Res_wqdpm_01. The time series shows that some of the constituents reach nearly steady-state conditions in the 30 days of simulation, but the phytoplankton and related particulate constituents are just beginning to increase at the end of the simulation time.

The spatial distribution of constituents, given in Res_wqdpm_01_SPO, shows that fairly strong stratification develops after 30 days and there are noticeable profiles of dissolved oxygen throughout the lake. The inflow dye indicates that the inflow is partially a bottom flow that reaches about halfway toward the dam in 30 days. The Algrt indicates that phytoplankton are growing in the epilimnion of the lake, and dying in the hypolimnion.

A plot of the velocity and dissolved oxygen distribution along the longitudinal–vertical slice is shown in Figure 7-4. The figure shows the inflow entering the lake as an interflow with a slight surface return current toward the inflow. The dissolved oxygen concentrations and stratification increase from the headwater toward the dam. There is a pocket of lower dissolved oxygen water developing along the bottom at about half the distance to the dam.

The time series results showing that many of the constituents including phytoplankton are just beginning to increase at the end of the simulation suggest that initializing all but temperature and dissolved oxygen to zero at the beginning of the simulation may be giving an incorrect result. The lake has a volume of $52.3 \times 10^6 \ m^3$ and at an inflow of 5 m^3/s has a residence time of 121 days, indicating that the inflow could not establish the proper initial distribution of constituents in the lake. This suggests one should run the simulation out to about 120 days, or three times as long as was done, to see if the results are more realistic. Another test on the response of the lake to inflows is to increase the inflow and outflow to the mean monthly flow through the lake that gives a 30 day residence time which is 15 m^3/s. Each should be tried to determine the effects on the results.

7.3 The Closed Loop Cooling Reservoir

The closed loop cooling reservoir is formed as done previously with a discharge of 107 m^3/s near the head of the reservoir, an underflow weir across the southern arm, and an intake from the southern arm. For the WQDPM application, the headwater inflow is left in and is removed at the dam. The input data is given in the example application folder Res_wqdpm_02. The headwater inflow has the same chemistry as the previous example. The facility has a 10° C temperature rise between the intake and discharge and a decrease in dissolved oxygen across the facility of 6.0 mg/l. The need for the latter will be explained later.

The problem again is how to initialize the reservoir. The 107 m^3/s circulating flow effectively reduces the residence time in the reservoir to approximately 6 days and may mix the headwater inflow quickly. The simulation is initialized the same as for the previous application for comparison. The time series results given in

Res_wqdpm_02_TSO tend to indicate that the higher rate of circulation does allow the reservoir to reach steady constituent concentrations faster than in the lake application.

The longitudinal–vertical profiles along the centerline of the lake given in Res_wqdpm_02_SPO show circulation and temperature distributions similar to those found in the Res_TSC_02 example application. The present application shows that the headwater virtual dye is mixed almost uniformly through the reservoir, and there is a relatively uniform distribution of the remaining water quality constituents. One notable exception is the dissolved oxygen, which shows a pocket of high dissolved oxygen in a region of relatively high Algrt. The circulation and the dissolved oxygen distribution is shown in Figure 7-5. The figure shows the two-layered circulation established in the cooling reservoir, and a complex dissolved oxygen distribution related to it and other processes.

The distribution of low dissolved oxygen water on the surface of the reservoir is related to the reduction of 6 mg/l of dissolved oxygen through the facility. An initial simulation without it resulted in highly supersaturated dissolved oxygen off the facility discharge. It was much higher than could be accounted for by poor choice of the water quality parameters used in the simulation. Usually, the dissolved oxygen for high temperature circulating water discharges is near saturation at the discharge temperature. The initial simulation required reducing the dissolved oxygen through the facility by at least 6 mg/l to maintain this condition.

7.4 An Example Lake Study Problem

An example water quality problem that uses actual meteorologic and inflow data is to make a preliminary estimate of how much the water quality changes between the inflow and release at the dam if the lake is built. The problem requires that analyses be performed during different times of the year. Given in Table 7-4 are monthly average meteorological data, inflow rates, and inflow water quality data taken from an inflow to a water body of a size roughly similar to the example lake. The required steps are as follows:

Use the meteorologic data to determine the monthly Cshe and Teq with the Teq Cshe Computation.exe routine.

Compute the Wx and Wy wind speed components using the wind speed and direction given in the data sheet. The direction is the azimuth measured from north from which the wind is blowing. For example, an azimuth of 120 degrees means that the wind is approximately out of the east southeast..

Set up a separate inflow data file for each month using the available flow data and water quality data.

Start with the first month and initialize the lake at the concentrations in the inflow.

Use the profiles of each of the dissolved and particulate constituents, and phytoplankton carbon at the end of the month to be the initial profile for the beginning of the next month. If a profile is specified for the location ipwest = ipeast = 12 and jpsouth = jpnorth = 9, a single profile can be printed out in the spatial output file for each constituent that can be transferred to a spreadsheet. If the profiles are lined up in the order required in the next month's input data initialization, the completed set of profiles can be transferred from the spreadsheet back into the next months input data table. Make sure that the number of layers, KM-2, does not change from month to month.

The inflow water chemistry has no values for particulate concentrations. One way to include the particulate constituents is to assume a particulate concentration that is a fraction of the dissolved concentration, and to reduce the latter proportionately.

Examination of the water chemistry shows that the NO_3 concentrations are much larger than the PO_4 concentrations. Examination of Equation 13.4.2 in Table 13-2, which shows the limitation of each of these constituents on phytoplankton growth, indicates that growth will be phosphorous-limited rather than nitrogen-limited.

Examination of the inflow rates shows that through the late winter and spring months the inflows are higher than the estimated 15 m^3/s that would theoretically flush the lake in 30 days, and in the summer the inflows are much less than this value. Relative to the water quality modeling, the lake is a flow-through lake in the late winter and spring and becomes a storage lake in the summer and fall. Use this fact to help interpret the results.

7.5 Application of the Water Quality Dissolved Particulate Model to Estuaries and Coastal Waters

The main difference when applying WQDPM or any water quality model to an estuary is the need to establish the concentration of constituents at the tidal or open boundary. These concentrations need to be specified so that the water quality of the tidal flow into the water body is known, but these are not always easy data to obtain. The second difference when applying one of the models to an estuary is the amount of data required to initialize the water quality profiles.

One way to let the estuarine water quality initialization and boundary conditions establish themselves relative to discharges within the estuary is to connect the estuary to a large harbor or coastal region. The estuary can freely exchange water with the different water quality constituents across its boundary. The bathymetry for such a configuration for a simple tidal canal connected to a coastal region is given in Table 7-5. The canal itself is 3 cells wide and 9 cells long. The coastal region is specified as an additional 9 cells wide and 9 cells long.

The project is set up as Can_WQDPM_01 and the details of the input data are given in its example applications folder. The project has an inflow at the head of the canal at a flow rate of 0.5 m^3/s, with all of the WQDPM dissolved constituents specified. The inflow includes phytoplankton to help initialize its density.

The tidal boundaries are set at the outer or eastern end of the coastal region for which the tidal elevations, boundary temperature, and boundary dissolved oxygen are specified. The canal and coastal region are initialized for temperature and dissolved oxygen, and a constituent dye is used to estimate residence times. Salinity is not initialized, allowing a baroclinic density flow to establish itself relative to the freshwater inflow. The relatively high velocities that initially occur near the eastern boundary due to the density flows require the use of a relatively small time step.

The objective of the simulation is to let the canal boundary response to the different water quality parameters of the inflow develop naturally rather than be forced at the boundary.

A second project, Can_WQDPM_02, is set up to illustrate the effects of constraining the canal tidal inflows and outflows on its water quality distribution resulting from the discharge.

7.5.1   Results for Project Can_WQDPM_01

The detailed results for the Can_WQDPM_01 project are given in its example application folder. The tidally averaged flow field and dissolved oxygen distribution within the canal is shown in Figure 7-6. The figure shows that the discharge establishes itself as a surface buoyant outflow on top of intruding coastal water. The dissolved oxygen is low near the discharge, is stratified along with the outflow, and approaches the tidal boundary value.

The detailed results in the example application folder show that no phytoplankton build up within the canal, and no particulate constituents are formed.

7.5.2   Results of Project Can_WQDPM_02

The second canal project is set up to illustrate the effects of constraining the circulation within the canal on its water quality due to the discharge. In this project a wave surge barrier is placed across the mouth of the canal extending down to a depth of about 3.5 m. The purpose of the barrier is to protect the land surrounding the canal for development.

The input data file for the project is found in its example application folder, and differs from the first project only by the inclusion of the barrier.

The resulting tidally averaged velocity and dissolved oxygen distribution within the canal is shown in Figure 7-7. The figure shows that the discharge within the canal is a buoyant outflow toward the barrier, but it circulates back and forth toward the bottom of the barrier. It also shows a net outflow under the barrier.

The dissolved oxygen within the canal is lower than for the first project. It increases from near the bottom of the barrier toward the bottom. The latter is due to the intrusion of higher dissolved oxygen water on incoming tides.

The detailed results given in the example applications folder for the tidally averaged spatial distributions show the relatively high velocity out of the canal under the surge barrier. Phytoplankton build up in the canal, and particulate biochemical oxygen demand (BOD_p) is produced.

7.5.3   Canal Flushing Times

The results for both canal projects indicate that there is a relatively short flushing time within the canal that might be related to the discharge itself or due to a combination of it and the salinity intrusion. The volume of the canal is approximately $5 \times 10^4 \, m^3$ and based on the discharge flow rate alone would have a flushing time of approximately 1 day. Tidal exchange makes this smaller.

For tidal exchange, the maximum rate of change of water surface elevation due to the tide is 1.0×10^{-4} m/s. Over the surface area of the canal this gives a maximum tidal inflow or outflow rate of approximately 1.2 m^3/s. The tide alone could reduce the flushing time to approximately 0.5 days. When combined with the discharge inflow, it reduces to approximately 0.3 days and corresponds closely with the values given for both projects.

Project Can_WQDPM_03 is run with no discharge into the canal and no tidal barrier. There is no induced baroclinic flow due to the stratified discharge. The tidally averaged velocities in the example application folder indicate that there is a net barotropic flow with small bottom inflows and an outward surface flow. The results of the residence time computations show that near the head of the canal it increases to 5 or 6 days. This indicates that the lower residence times in the first two projects are due to the discharge and mixing resulting from the tidal flow.

7.5.4   Suggested Study Problems

Do the following suggested study problems for both Can_WQDPM_1 and Can_WQDPM_2 to compare the results:

The effects of the canal itself on water quality can be incorporated in the simulations by specifying a measured SODm in the WQDPM parameter file. Do this for all three projects to determine its impact on dissolved oxygen. The release of NH_4 from the sediment is proportional to sediment oxygen demand, and it should increase with the additional SODm.

It appears that the results with the discharge may be sensitive to its specified phytoplankton concentration. For the first two projects, try gradually increasing values of phytoplankton concentrations to determine the changes.

Specify dissolved water quality constituent and phytoplankton concentrations uniformly at the tidal boundary as background conditions. Place a surface wind from south to north that may help mix the coastal region without disturbing the circulation within the canal. Include the discharge into the canal, but without phytoplankton. Determine the combined effects of background conditions and discharge conditions on the water quality of the canal.

Go back to the estuary project examined in Chapter 4 and extend its bathymetry to include a coastal region. Initialize the salinity profile and tides at the coastal region tidal boundary. Determine the differences in circulation and salinity distributions when using a coastal region versus specifying their value right at the mouth of the estuary.

7.6 Density Dependent Phytoplankton Grazing

The WQDPM has an option for phytoplankton grazing to be density dependent. This means that the zooplankton grazing rate is a function of the phytoplankton density, making grazing dependent on the square of the phytoplankton density. Density-dependent grazing results in phytoplankton densities, and hence the concentrations of the water quality constituents related to it approach values that change little with time for constant inflow concentrations and other parameters. Additional background on density-dependent grazing is presented in Gentleman et al. (2000) and its formulation in WQDPM is presented in Edinger et al. (2001).

The density-dependent grazing option is applied by setting the Phyt_yes flag in the_WQDPM.dat file as shown in Table 3-5 to a value of 2. The density-dependent grazing rate is then controlled by the value set for gzoo shown in Table 3-5. For the density-dependent computations, gzoo ranges from 10 to 100.

Two example applications are given to show the effects of including density-dependent grazing in WQPDM. The first is a set of tank tests included in folder 7.01a, called Tank Grazing Tests. These include time series diagrams for each case. The second example application is for a reservoir given in folder 7.06a wqdpm_01a. Its results can be compared to the case given in folder 7.06 Res_wqdpm_01. The time series file in Res_wqdpm_01 shows the phytoplankton density continually increasing with time through the simulation. However, with a phytoplankton density-dependent zooplankton grazing rate, the time series file in Res_wqdpm_01a shows that the phytoplankton density reaches a constant value fairly quickly into the simulation, as do the time series of the remaining water quality parameters.

Each of the example applications in this chapter should be re-run for a case including density dependent grazing for comparison to the previous results. The additional cases can be run by copying the folder containing the existing case to be run and renaming it, setting the phytoplankton option to Phyto_yes =2 and gzoo = 50. Place an Abcontrolfiles.dat into the new folder and run it using DMAWin.exe. A detailed description for using the latter is in the "Programs and folders in the INTROGLLVHT Model folder" contained in the INTROGLLVHT Model software folder.

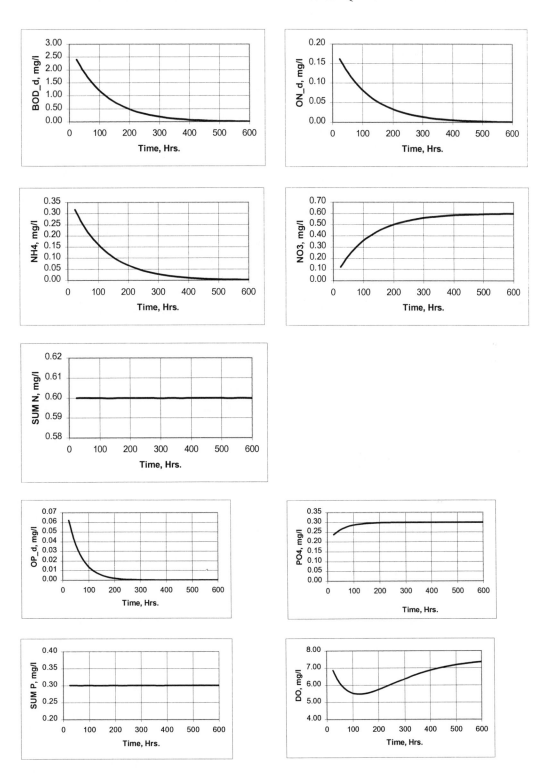

Figure 7-1. Results of tank test simulations with Water Quality Dissolved Particulate Model (WQDPM) for project Tank_WQDPM_01a. Dissolved constituents only, no phytoplankton or sediment exchange. Notation in abbreviations.

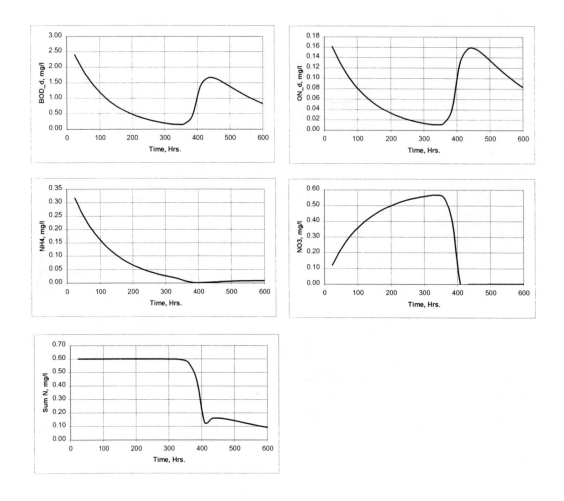

Figure 7-2. Tank_WQDPM_01c. Notation in abbreviations.

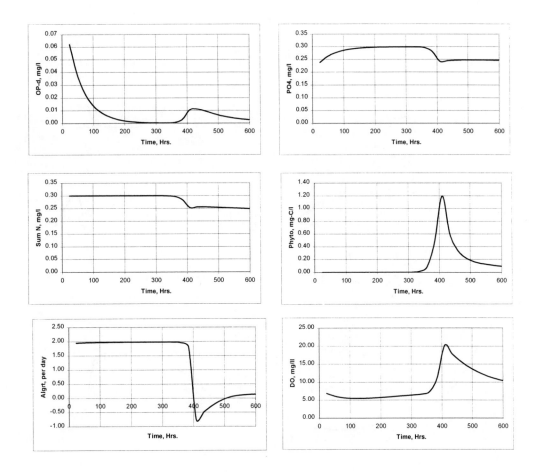

Figure 7-2 (continued)

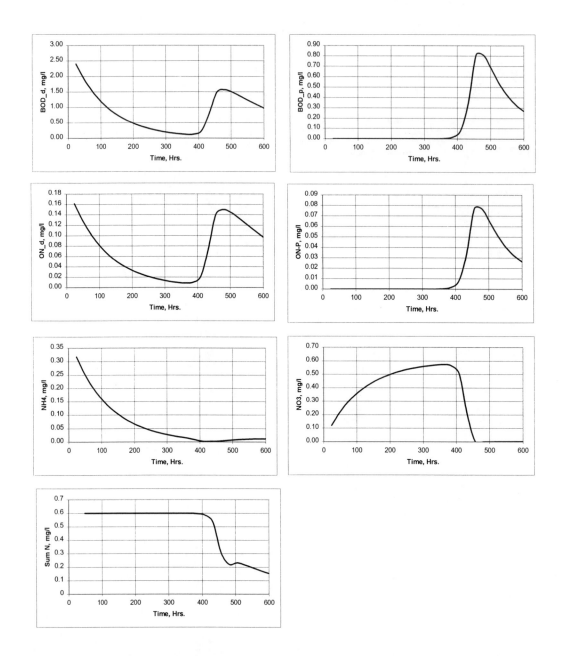

Figure 7-3. Results of tank test simulations with WQDPM for project Tank_WQDPM_01e. Dissolved and particulate constituents with phytoplankton, settling and sediment exchange. Notation in abbreviations.

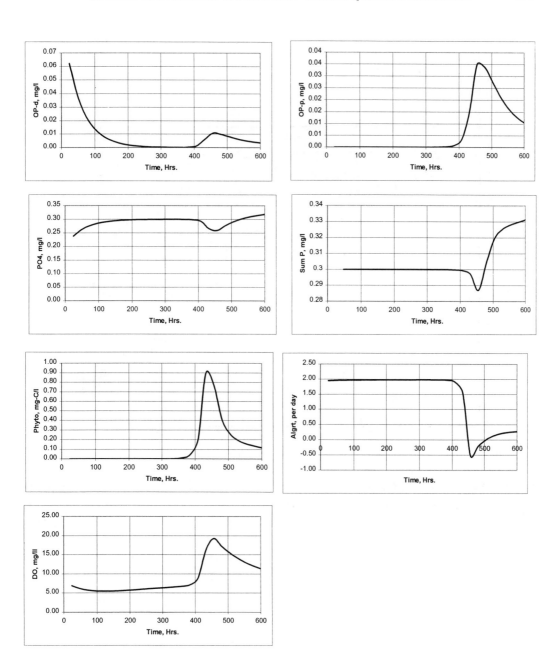

Figure 7-3 (continued)

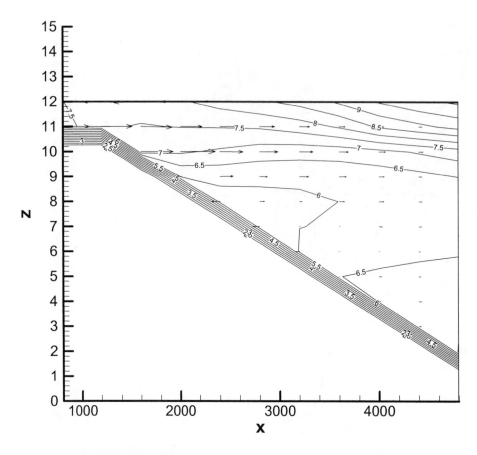

Figure 7-4. Longitudinal-vertical profile through lake of velocities and dissolved oxygen from application Project Res_WQDPM_01.

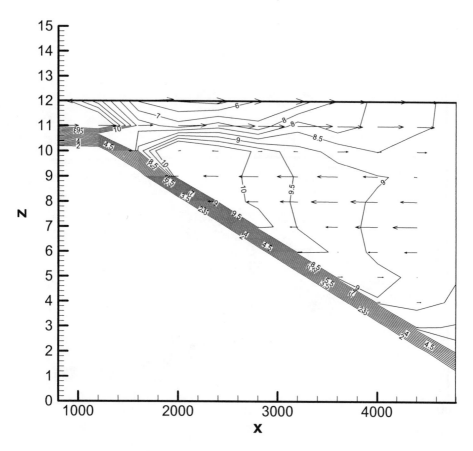

Figure 7-5. Longitudinal-vertical profile through lake of velocities and dissolved oxygen from application project Res_WQDPM_02.

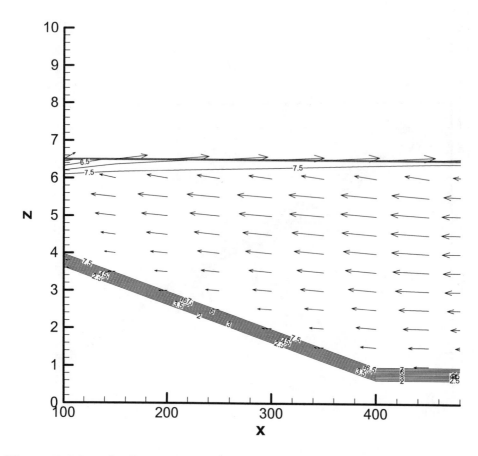

Figure 7-6. Longitudinal-vertical velocity and dissolved oxygen distribution through canal for project Can_WQDPM_01.

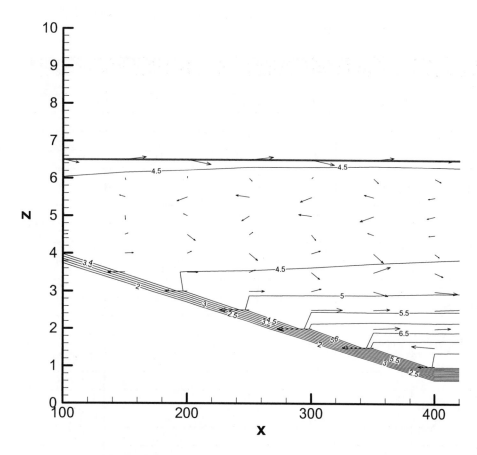

Figure 7-7. Longitudinal-vertical velocity and dissolved oxygen distribution through canal for project Can_WQDPM_01.

Table 7-1. Water quality dissolved particulate model parameters used for each tank test using Tank_WQDPM_01a,b,c,d,e.

Parameter	Case 01a	Case 01b	Case 01c	Case 01d	Case 01e
NH4_yes	1	1	1	1	1
K12	0.20	0.20	0.20	0.20	0.20
OND_yes	1	1	1	1	1
K71	0.10	0.10	0.10	0.10	0.10
ONP_yes	0	0	0	1	1
Vs7	0.0	0.0	0.0	0.0	0.2
NO3_yes	1	1	1	1	1
K2d	0.0	0.0	0.0	0.0	0.0
PO4_yes	0	1	1	1	1
BODD_yes	1	1	1	1	1
kd	0.15	0.15	0.15	0.15	0.15
BODP_yes	0	0	0	1	1
Vs5	0.0	0.0	0.0	0.0	0.2
DO_yes	1	1	1	1	1
OPD_yes	0	1	1	1	1
K83	0.0	0.22	0.22	0.22	0.22
OPP_yes	0	0	0	1	1
Vs8	0.0	0.0	0.0	0.0	0.08
Phyt_yes	0	0	1	1	1
K1c	0	0	1.6	1.6	1.6
K1d	0	0	0.5	0.5	0.5
K1r	0	0	0.07	0.07	0.07
Vs4	0	0	0.0	0.0	0.3
Is	90.0	90.0	90.0	90.0	90.0
As	0.70	0.70	0.70	0.70	0.70
fe	0.10	0.10	0.10	0.10	0.10

Table 7-1 (continued)

Parameter	Case 01a	Case 01b	Case 01c	Case 01d	Case 01e
Sediment					
S4	0.0	0.0	0.0	0.0	0.09
SP5	0.0	0.0	0.0	0.0	0.07
SP7	0.0	0.0	0.0	0.0	0.09
Spnh3	0.45	0.45	0.45	0.45	0.45
SODm	0.0	0.0	0.0	0.0	0.0
SedNH3m	0.0	0.0	0.0	0.0	0.0
SedPO4m	1.5	1.5	1.5	1.5	1.5
Environ-mental Parameters.					
Hsc	150.0	150.0	150.0	150.0	150.0
Cloud Cover	2.0	2.0	2.0	2.0	2.0
Wad	3.6	3.6	3.6	3.6	3.6

Table 7-2. Tank_WQDPM_01_INP.dat used for test tank simulations.

```
Tank_WQDPM_01
 $1.nwqm              3
 $2.Inflow Conditions
 $ninflows            0
 $qinflow,iinflow,jinflow,kinflow
 $intake,inintake,jintake,kintake
 $temp,saln,const,cbod,on,nh3,do(d),no3,op,po4,phyt,cbod_p,on_p,op_p
 $.4 Outflow Conditions
 $noutflows           0
 $qoutflow,ioutflow,joutflow,koutflow
 $4.Elevation Boundary Conditions
 $nelevation  kts            0              2
 $iewest,ieeast,jesouth,jenorth
 zmean,zamp,tmelag,tideper
 k,temp,saln,const,cbod,on,nh3,do(d),no3,op,po4,phyt,cbod_p,on_p,op_p for k=2,km-2
 $5.Initialize Water Quality Profiles
 $ninitial          1
 k,temp,saln,const,cbod,on,nh3,do(d),no3,op,po4,phyt,cbod_p,on_p,op_p for k=2,km-2
      2    30.00   0.00   0.00    3.00   0.20   0.4  8.0  0.0  0.10  0.20  0.0  0.0
 0.0  0.0
      3    30.00   0.00   0.00    3.00   0.20   0.4  8.0  0.0  0.10  0.20  0.0  0.0
 0.0  0.0
 $6.External Parameters
 Chezy, Wx,Wy,CSHE, TEQ, Rdecay, Lat.
     35.00     3.00      0.00     0.00   30.00     0.00      0.00
 $7.Output Profiles
 $nprofiles           0
 $ipwest,ipest,jpsouth,jpnorth
 $u-vel, v-vel, w-vel
 $nconstituents
 $I-const(1),I-const(2), I-const(3), etc
 $8.Output Surfaces
 $nsurfaces  $nconstituents         0              0
 $U-vel   V-vel   0.0000000E+00   0.0000000E+00
 $I-const(1), I-const(2), I-const(3), etc.
 $9.Output Time Series
 $ntimser            13
 $nconst, iconst, jconst,kconst
      4      2      2      2
      5      2      2      2
      6      2      2      2
      7      2      2      2
      8      2      2      2
      9      2      2      2
     10      2      2      2
     11      2      2      2
     12      2      2      2
     13      2      2      2
     14      2      2      2
     15      2      2      2
     15      2      2      3
 $10.Simulation time conditions
 $dtm  tmend   200.0000         960.0
 $tmeout  tmeserout   960.0        24.0
 $11.Internal Boundary Locations
 $nintbnd            0
 $ibwest,ibeast,jbsouth,jbnorth,ktop,kbottom
 $12. Constituent Averages
 $nconarv            0
 $nconstarvs
 $13. Groundwater Inflow
 $ngrndwtr           0
```

Table 7-3. Input data for project application Res_WQDPM_01 with inflow and outflow to a lake.

```
Res_wqdpm_01
 $1.nwqm          3
 $2.Inflow Conditions
 $ninflows            1
 $qinflow,iinflow,jinflow,kinflow
 $intake,inintake,jintake,kintake
 $temp,saln,const,cbod,on,nh3,do(d),no3,op,po4,phyt,cbod_p,on_p,op_p
     5.00    2    8    2
     0    0   0    0
    20.00     0.00  100.00   0.5   1.5   1.0    8.0   2.0  1.0  2.0  0.0 0.0 0.0 0.0
 $.4 Outflow Conditions
 $noutflows           1
 $qoutflow,ioutflow,joutflow,koutflow
     5.00    12    9    11
 $4.Elevation Boundary Conditions
 $nelevation  kts          0             2
 $iewest,ieeast,jesouth,jenorth
 zmean,zamp,tmelag,tideper
 k,temp,saln,const,cbod,on,nh3,do(d),no3,op,po4,phyt,cbod_p,on_p,op_p for k=2,km-2
 $5.Initialize Water Quality Profiles
 $ninitial         1
 k,temp,saln,const,cbod,on,nh3,do(d),no3,op,po4,phyt,cbod_p,on_p,op_p for k=2,km-2
      2     20.00     0.0    0.0   0.0   0.0  0.0    8.0   0.0  0.0  0.0  0.000 0.0 0.0 0.0
      3     20.00     0.0    0.0   0.0   0.0  0.0    8.0   0.0  0.0  0.0  0.000 0.0 0.0 0.0
      4     20.00     0.0    0.0   0.0   0.0  0.0    8.0   0.0  0.0  0.0  0.000 0.0 0.0 0.0
      5     20.00     0.0    0.0   0.0   0.0  0.0    8.0   0.0  0.0  0.0  0.000 0.0 0.0 0.0
      6     20.00     0.0    0.0   0.0   0.0  0.0    8.0   0.0  0.0  0.0  0.000 0.0 0.0 0.0
      7     20.00     0.0    0.0   0.0   0.0  0.0    8.0   0.0  0.0  0.0  0.000 0.0 0.0 0.0
      8     20.00     0.0    0.0   0.0   0.0  0.0    8.0   0.0  0.0  0.0  0.000 0.0 0.0 0.0
      9     20.00     0.0    0.0   0.0   0.0  0.0    8.0   0.0  0.0  0.0  0.000 0.0 0.0 0.0
     10     20.00     0.0    0.0   0.0   0.0  0.0    8.0   0.0  0.0  0.0  0.000 0.0 0.0 0.0
     11     20.00     0.0    0.0   0.0   0.0  0.0    8.0   0.0  0.0  0.0  0.000 0.0 0.0 0.0
     12     20.00     0.0    0.0   0.0   0.0  0.0    8.0   0.0  0.0  0.0  0.000 0.0 0.0 0.0
     13     20.00     0.0    0.0   0.0   0.0  0.0    8.0   0.0  0.0  0.0  0.000 0.0 0.0 0.0
 $6.External Parameters
 Chezy, Wx,Wy,CSHE, TEQ, Rdecay, Lat.
    35.00     0.00     0.00   25.00    35.00    0.00    28.20
 $7.Output Profiles
 $nprofiles           1
 $ipwest,ipest,jpsouth,jpnorth
 $u-vel, v-vel, w-vel
 $nconstituents
 $I-const(1),I-const(2), I-const(3), etc
      2    12    8    8
      1    0    0
     14
      1    3    4    5    6    7    8    9    10    11 12 13 14   15
 $8.Output Surfaces
 $nsurfaces  $nconstituents         0              0
 $U-vel  V-vel   1.0000000E+00   0.0000000E+00
 $I-const(1), I-const(2), I-const(3), etc.
 $9.Output Time Series
 $ntimser           14
 $nconst, iconst, jconst,kconst
      1     7    8    3
      4     7    8    3
      5     7    8    3
      6     7    8    3
      8     7    8    3
      7     7    8    3
      8     7    8    3
      9     7    8    3
     10     7    8    3
     11     7    8    3
     12     7    8    3
     13     7    8    3
     14     7    8    3
     15     7    8    3
 $10.Simulation time conditions
 $dtm  tmend    240.0000        720.5
 $tmeout  tmeserout    720.0    12.0
 $11.Internal Boundary Locations
 $nintbnd         0
 $ibwest,ibeast,jbsouth,jbnorth,ktop,kbottom
 $12. Constituent Averages
 $nconarv          0
 $nconstarvs
 $13. Groundwater Inflow
 $ngrndwtr          0
```

Table 7-4. Meteorological and river inflow data for example lake by month (latitude of lake is 46.6° North).

	Month	Jan	Feb	Mar	Apr	May	June	July	August	Sept
Cloud Cover	Tenths	8.2	8.1	8.9	7.7	7.9	7.9	6.4	6.1	6.4
Dew Point Temp	C	1.2	1.84	3.4	3.8	8.8	10.1	11.1	12.7	11.1
Wind speed	m/s	1.52	1.49	2.26	1.61	1.39	1.46	1.4	1.15	1.51
Wind direction	Az. Deg.	122	113	159	129	108	126	116	93	103
River flow	cu.m/s	35.6	23.5	30.0	14.7	9.7	8.1	5.2	4.2	4.5
Water Temp.	C	6.8	7.1	8.6	10.7	14.1	16.3	17.5	18.1	17.5
BOD_d	mg/l	15.5	11.5	21.5	11.1	16.5	15.1	17.2	18.6	16.9
ON-d	mg/l	0.01	0.01	0.01	0.01	0.01	0.01	0.01	0.01	0.01
NH4	mg/l	0.02	0.01	0.01	0.01	0.02	0.01	0.02	0.01	0.02
DO	mg/l	11.1	11.1	10.6	10.1	10.1	9.1	8.6	8.2	9.6
NO3	mg/l	0.53	0.59	0.51	0.55	0.46	0.42	0.36	0.3	0.24
OP_d	mg/l	0.005	0.006	0.005	0.006	0.004	0.005	0.004	0.005	0.004
PO4	mg/l	0.018	0.018	0.016	0.02	0.017	0.012	0.012	0.011	0.011
Phyto	mg/l									
BOD_p	mg/l									
ON_p	mg/l									
OP-p	mg/l									

Table 7-5. Bathymetry data for a tidal canal connected to a coastal region.

```
CAN_BATH_01
        20    IM
        11    JM
        16    KM
   50.00000   DX
    8.000000  DY
    0.500000  DZ
        2    3    4    5    6    7    8    9   10   11   12   13   14   15   16   17   18   19
  10  0.0  0.0  0.0  0.0  0.0  0.0  0.0  0.0  0.0  6.0  6.0  6.0  6.0  6.0  6.0  6.0  6.0  6.0
   9  0.0  0.0  0.0  0.0  0.0  0.0  0.0  0.0  0.0  6.0  6.0  6.0  6.0  6.0  6.0  6.0  6.0  6.0
   8  0.0  0.0  0.0  0.0  0.0  0.0  0.0  0.0  0.0  6.0  6.0  6.0  6.0  6.0  6.0  6.0  6.0  6.0
   7  3.0  3.5  4.0  4.5  5.0  5.5  6.0  6.0  6.0  6.0  6.0  6.0  6.0  6.0  6.0  6.0  6.0  6.0
   6  3.0  3.5  4.0  4.5  5.0  5.5  6.0  6.0  6.0  6.0  6.0  6.0  6.0  6.0  6.0  6.0  6.0  6.0
   5  3.0  3.5  4.0  4.5  5.0  5.5  6.0  6.0  6.0  6.0  6.0  6.0  6.0  6.0  6.0  6.0  6.0  6.0
   4  0.0  0.0  0.0  0.0  0.0  0.0  0.0  0.0  0.0  6.0  6.0  6.0  6.0  6.0  6.0  6.0  6.0  6.0
   3  0.0  0.0  0.0  0.0  0.0  0.0  0.0  0.0  0.0  6.0  6.0  6.0  6.0  6.0  6.0  6.0  6.0  6.0
   2  0.0  0.0  0.0  0.0  0.0  0.0  0.0  0.0  0.0  6.0  6.0  6.0  6.0  6.0  6.0  6.0  6.0  6.0
      -999    geomdelim
```

8. APPLICATION OF THE SEDIMENT SCOUR AND DEPOSITION MODEL

The sediment model, SED, computes the net rate of scour/deposition for each location within a water body. The SED model is derived in Chapter 14, and the names of the parameters used in the SED model, their default values, and their definitions are summarized in Table 8-1. The SED output variables are the bottom sediment concentration (Sed), the cumulative scour/deposition rate over the simulation period (Sedrt), and the instantaneous scour/deposition rate (Isrdt) at the time of output to the spatial output (SPO) or time series output (TSO) file. Each is defined in Chapter 14.

At least the average sediment rate, Sedrt, should be printed out. For steady-state cases, it is often useful as well to print out the instantaneous sediment rate, Isdrt, to make sure that the simulation time is long enough for Sedrt to become equal to Isdrt. In tidal cases, it is often useful to print out Isdrt, along with Sedrt, to determine the variability of the latter over a tidal cycle.

In this chapter, the SED scour/deposition model is first set up for a simple river channel with a jetty. It is then set up as a tidal river to illustrate the effects of tidal currents in varying the instantaneous scour/deposition rate.

A project application is then made to an open coastal region. This is first run with no breakwater to determine the scour/deposition pattern due to unhindered tidal currents off the beach. It is then run as a project with a harbor breakwater.

8.1 Application of the Sediment Scour and Deposition Model to a River

The Sed model is first set up for a simple river reach that has the bathymetry shown in Table 8-2. The river is 11 segments wide and 19 segments long. It has a uniform depth of 3 m sliced into 1 m layers.

8.1.1 The Sediment Scour and Deposition Model Input File

The input file for the river SED simulation is given in Table 8-3. The project application is named Riv_Sed_01. The nwqm is set to 4 to indicate that the SED model is being run. There are 11 inflows, one for each cell at the head of the channel. Each inflow is set at 3.0 m^3/s and goes into the top layer. None of the inflows are connected to an intake. Each inflow has a temperature of 30° C, zero salinity, and zero concentration of suspended sediment.

The downstream outflows are accounted for by specifying a downstream elevation boundary. The downstream elevation boundary for river cases allows a variable lateral distribution of velocity that results from upriver disturbances such as variable bathymetry or jetties to approach the boundary. It also allows for different combinations of inflows and outflows to be specified along the river without having to adjust outflows at the downstream boundary.

The fixed elevation boundary at the downstream end (I = 20) extends from J = 2 to J = 12. The boundary mean elevation, amplitude, tidal lag, and tidal period are all set to zero. A temperature of 30° C, a salinity of zero, and a suspended sediment concentration of zero are specified for each layer at the downstream elevation boundary. The temperatures throughout the river channel are initialized to 30° C, and the salinity and suspended sediment concentration are initialized to zero.

When the sediment model is set up, the nsurfaces is set to 1 and the bottom velocity components, the bottom suspended sediment concentrations, the average sediment scour/deposition rate, and the instantaneous scour/deposition rate can be printed out. The nconstits is set to the number of the latter that are desired, and their constituent numbers are chosen from Table 8-1.

A time series file is specified for the bottom suspended sediment concentration and sediment scour/deposition rate at I = 10, J = 10, and K = 4. The constituent numbers for the suspended sediment concentration, the cumulative scour/deposition rate, and the instantaneous scour/deposition rate for the time series files are the same as in Table 8-1. Also, time series of elevation and the U and V velocity components can be printed out using the constituent numbers given in Table 8-3 (see also Table 2-3).

An internal boundary representing the breakwater is located at I = 10 and extends part of the way across the river from J = 2 to J = 6. It extends from the surface to the bottom. Average concentrations are not printed out for any parameter.

8.1.2   Specification of Sediment Scour and Deposition Rate Parameters

The SED rate parameters were discussed in Chapter 3, Section 3.3.4, and the default values are saved in the INTROGLLVHT Model file SED_WQM.dat. The name of the project water quality parameter model should be Riv_SED_01_WQM.dat. A skeleton file by that name is automatically generated if the INTROGLLVHT Input File.exe routine is used to set up the input file.

The default sediment rate parameters need to be copied from the file SED_WQM.dat and placed in the project Riv_SED_01_WQM.dat file. The project parameters can then be changed in the latter without disturbing the default values.

8.1.3   Simulation Results

The river SED simulations are run for the parameter values given in the Riv_SED_01_WQM.dat file. The horizontal distribution of bottom velocity components and the distribution of the average sediment scour/deposition rates are shown in Figure 8-1.

In Figure 8-1, the velocity field shows that an eddy establishes behind the breakwater. The scour/deposition rate, in g/m²/yr, is negative and large off of and downstream from the breakwater. The negative value indicates that scouring is taking place. Some of the material scoured from this region is deposited downstream and the rest flows out at the downstream open boundary.

As can be seen from the Riv_SED_01_SPO.dat table in the example project folder, the instantaneous scour/deposition rates are nearly equal to the average scour/deposition rates, indicating that steady-state conditions are almost reached by the end of the simulation. Similarly, the Riv_SED_01_TSO.dat time series table shows that the scour/deposition rate is still building up slowly at the end of the simulation but is close to reaching steady-state conditions. The bottom sediment concentrations are too small to show up in the SPO table, but their magnitude can be found in the TSO table.

8.1.4   *Additional Study Problems*

Following are some additional study problems.

Riv_SED_01a. Determine how the scour/deposition pattern varies for different channel velocities. Adjust the velocities by varying the dy in the bathymetry file.

Riv_SED_01b. Referring to Chapter 10 (which gives a range of parameters), perform simulations for different combinations of sediment parameters. In particular, try different particle diameters.

Riv_SED_01c. Place different concentrations of suspended sediment in the river inflow to see how the scour/deposition varies.

Riv_SED_01d. Set up a river bathymetry with a more variable cross-section of depths including overbank areas along either shore and a channel down the middle. Determine how this changes the scour/deposition pattern.

Riv_SED_01e. Both the settling velocity and the scour rate depend on the density and viscosity of the river. Place a heated discharge with a temperature of 40° C and a flow rate of 10 m^3/s in the bottom layer at I = 2 and J = 2 to determine its influence on the scour/deposition pattern. Run the simulation first with a 30° C discharge temperature to determine the effect of the increased flow to the river channel.

8.2 Application of Sediment Scour and Deposition to a Tidal River

In this section, the SED scour/deposition model is applied to a tidal river to illustrate how the instantaneous scour/deposition rate varies with time due to a tidally varying bottom current, and to compare the resulting tidally averaged current pattern and scour/deposition rate with the nontidal river project. The steady river flow project, Riv_SED_01, is set up as a tidal river project named Riv_SED_02. The major change from the first project is that a tide with an amplitude of 0.6 m and a period of 12.45 hours is imposed at the downriver boundary.

The spatial output is set to obtain the bottom velocity components, and the average and instantaneous scour/deposition rate throughout the last two tidal cycles. The time series output is set to obtain the velocity component, average scour/deposition rate (sedrt), and instantaneous scour/deposition rate (isdrt) throughout a tidal cycle. The detailed output files can be found in the project example applications folder named Riv_SED_02.

8.2.1   Tidal River Scour Deposition Project Results

The time series of velocity and scour/deposition rate at a location off the breakwater are shown in Figure 8-2. The figure shows that scouring is highest at the time of maximum outgoing velocity, and there is some redeposition during the incoming tide. The time series output results given in the project folder show that the average scour/deposition rate is established quite rapidly into the simulation at about -85 $g/m^2/yr$, and this appears to be the average value of the isdrt in Figure 8-2.

The tidally averaged current and sedrt distribution are shown in Figure 8-3. The figure shows that the scouring rate off the breakwater is larger than for the nontidal river project, and that there is deposition behind the breakwater. The eddy behind the breakwater is quite small.

8.2.2   Additional Suggested Study Examples

Following are some additional suggested study examples.

Riv_SED_02a. Gradually decrease the tidal amplitude for a number of different simulations so that the tidal currents decrease. Discuss how the scour/deposition pattern approaches the nontidal river case.

Riv_SED_02b. Change the density throughout the river using salinity intrusion from the tidal boundary. Place a uniform salinity profile of 20 ppt at the tidal boundary and adjust the tidal mean height and amplitude until there is significant salinity intrusion up the river. Determine how the change in the flow field and water density influences the scour/deposition pattern.

Riv_SED_02c. Vary the salinity laterally across the tidal boundary from lower values at its northern end to higher values at its southern end by breaking the boundary up into groups of two cells each and changing the salinity for every two cells. Test the circulation induced for the lateral variation in salinity first with no jetty and no river flow, then for increasing river flows. Include the jetty to determine the effects of this kind of circulation on the sediment scour/deposition.

Res_SED_01. Perform an analysis of sediment scour/deposition for the cooling lake given in project Res_TSC_02. Examine the effects of the intake skimmer wall on bottom scour/deposition. Also, examine what happens in the discharge region as suspended sediment is recirculated from the intake area to the discharge area.

8.3 Application of the Scour/Deposition Model to a Coastal Water Project

Here, the scour/deposition model is applied in the coastal water project Coastal_Sed_01. The purpose of this project is to determine the scour patterns that exist due to the naturally occuring tidal currents. The project is set up for the same coastal bathymetry and time-lagged tidal boundaries as used in the project Coastal_brine_discharge_02. The SED_WQM.dat default parameters are used for the project.

8.3.1   Results of the Application of Sediment Scour and Deposition to a Tidal Coastal Regime

The bottom U-velocity component time series is given for two locations in Coastal_Sed_01_TS0. Examination of these data indicates that stationary-state conditions are reached within the time of simulation of 120 hours.

The sediment scour/deposition pattern is presented in Figure 8-4. The results indicate that scouring takes place from the beach outward, and is greater in the deeper water. It shows that there may be some distortion of the naturally occurring scour field due to the proximity of the boundaries.

8.3.2   Example Study Applications

Following are some example study applications.

Coastal_Sed_01a. Use the expanded grid setup for study example CBD_01a to determine if moving the tidal boundaries from the area of interest changes the sediment scour/deposition pattern.

Coastal_Sed_01b. Determine the sensitivity of the scour/deposition pattern to parameters in the _WQM.dat file. In particular, determine the sensitivity to particle diameter. Also, determine the sensitivity to water temperature.

8.4 Application of Sediment Scour and Deposition to a Coastal Water Project with a Marina Breakwater

A marina extending off the beach is incorporated in application project Coastal_Sed_02 for the same input conditions as used in project Coastal_Sed_01.

8.4.1   Results of Coastal Scour/Deposition with Marina

The results are shown in Figure 8-5. They indicate that the marina has a dramatic effect on the tidally averaged current and the scour/deposition pattern. An eddy gets established within the marina, and there is some net deposition within it. Scour takes place offshore of the marina, and extends up and down the coast from it.

8.4.2   Example Study Applications

Following are some example study applications.

Coastal_Sed_02a. Determine how the current pattern and scour/deposition pattern and intensity changes if the size of the marina is changed.

Coastal_Sed_02b. Determine how the current pattern and scour/deposition pattern and intensity changes with changes in properties of the sediment, particularly the particle size.

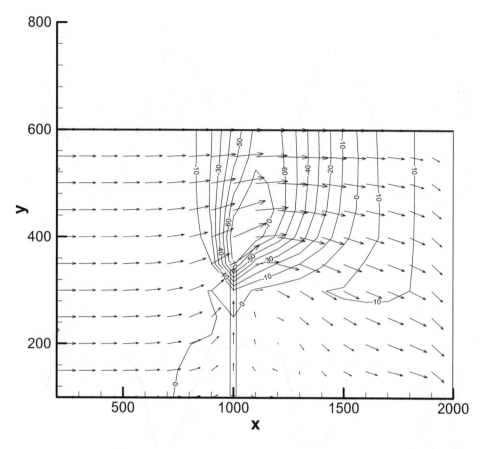

Figure 8-1. Scour/deposition pattern in a river with a breakwater for project Riv_SED_01. Scour/depositon rates are in g/m^2/yr.

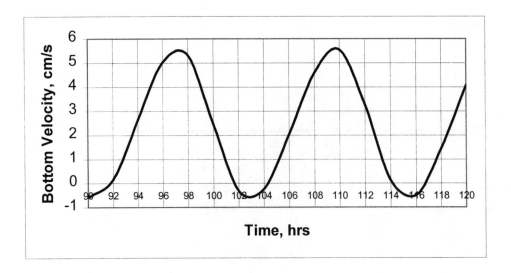

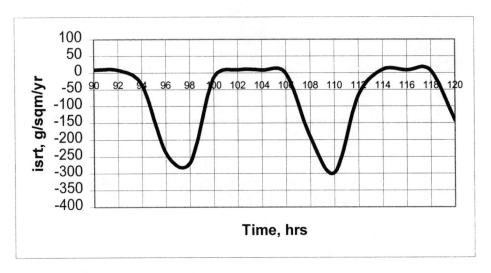

Figure 8-2. Time series of bottom velocity and instantaneous scour/deposition rate near end of breakwater for project Riv_Sed_02.

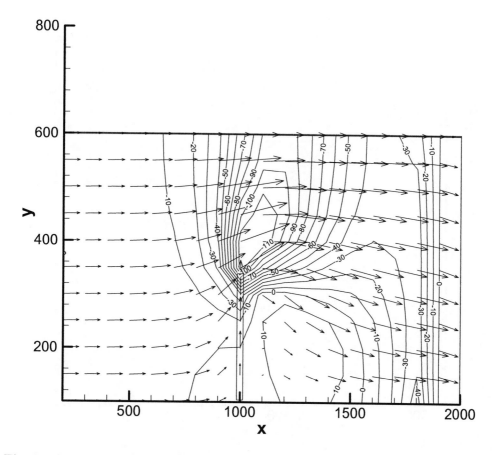

Figure 8-3. Scour/deposition pattern in a tidal river for project Riv_SED_02. Scour/deposition rates are given in $g/m^2/yr$.

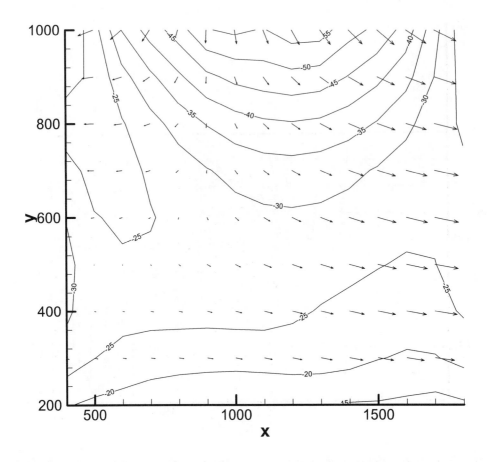

Figure 8-4. Scour/deposition pattern in a portion of coastline for project Coastal_Sed_01. Scour/deposition rates are gGiven in g/m^2/yr.

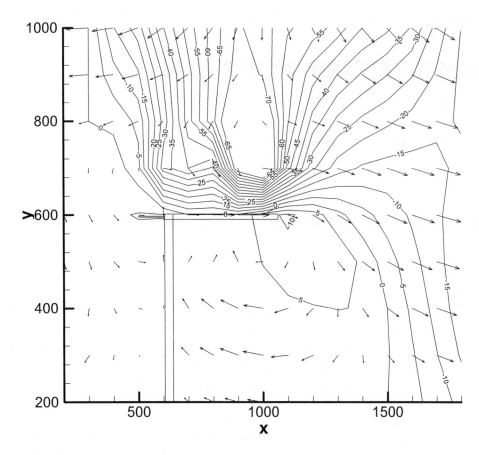

Figure 8-5. Scour/deposition pattern in a portion of coastline for project Coastal_Sed_02 with marina. Scour/deposition rates are given in g/m^2/yr.

Table 8-1. Constituents and parameters, their number, units, and definitions, as used in the sediment scour and deposition model.

Symbol	No.	Description
Temp.	1	Celsius
Salinity	2	ppt
Sed	3	Bottom sediment concentration, mg/l
Sedrt	4	Cumulative sediment rate, $g/m^2/yr$
Isdrt	5	Instanteous sediment rate, $g/m^2/yr$

Table 8-2. River bathymetry for sediment scour/deposition modeling.

```
River
          21     IM
          13     JM
           6     KM
    100.0000     DX
    50.00000     DY
    1.000000     DZ
               2    3    4    5    6    7    8    9   10   11   12   13   14   15   16   17   18   19
20
       12  3.0  3.0  3.0  3.0  3.0  3.0  3.0  3.0  3.0  3.0  3.0  3.0  3.0  3.0  3.0  3.0  3.0  3.0
3.0
       11  3.0  3.0  3.0  3.0  3.0  3.0  3.0  3.0  3.0  3.0  3.0  3.0  3.0  3.0  3.0  3.0  3.0  3.0
3.0
       10  3.0  3.0  3.0  3.0  3.0  3.0  3.0  3.0  3.0  3.0  3.0  3.0  3.0  3.0  3.0  3.0  3.0  3.0
3.0
        9  3.0  3.0  3.0  3.0  3.0  3.0  3.0  3.0  3.0  3.0  3.0  3.0  3.0  3.0  3.0  3.0  3.0  3.0
3.0
        8  3.0  3.0  3.0  3.0  3.0  3.0  3.0  3.0  3.0  3.0  3.0  3.0  3.0  3.0  3.0  3.0  3.0  3.0
3.0
        7  3.0  3.0  3.0  3.0  3.0  3.0  3.0  3.0  3.0  3.0  3.0  3.0  3.0  3.0  3.0  3.0  3.0  3.0
3.0
        6  3.0  3.0  3.0  3.0  3.0  3.0  3.0  3.0  3.0  3.0  3.0  3.0  3.0  3.0  3.0  3.0  3.0  3.0
3.0
        5  3.0  3.0  3.0  3.0  3.0  3.0  3.0  3.0  3.0  3.0  3.0  3.0  3.0  3.0  3.0  3.0  3.0  3.0
3.0
        4  3.0  3.0  3.0  3.0  3.0  3.0  3.0  3.0  3.0  3.0  3.0  3.0  3.0  3.0  3.0  3.0  3.0  3.0
3.0
        3  3.0  3.0  3.0  3.0  3.0  3.0  3.0  3.0  3.0  3.0  3.0  3.0  3.0  3.0  3.0  3.0  3.0  3.0
3.0
        2  3.0  3.0  3.0  3.0  3.0  3.0  3.0  3.0  3.0  3.0  3.0  3.0  3.0  3.0  3.0  3.0  3.0  3.0
3.0
     -999  geomdelim
```

Table 8-3. Input table for sediment scour/deposition project Riv_SED_01.

```
Riv_Sed_01
 $1.nwqm              4
 $2.Inflow Conditions
 $ninflows              11
 $qinflow,iinflow,jinflow,kinflow
 $intake,inintake,jintake,kintake
 $temp,saln,const,cbod,on,nh3,do(d),no3,op,po4,phyt
     3.00        2       2       2
     0       0       0       0
    30.00        0.00       0.00
     3.00        2       3       2
     0       0       0       0
    30.00        0.00       0.00
     3.00        2       4       2
     0       0       0       0
    30.00        0.00       0.00
     3.00        2       5       2
     0       0       0       0
    30.00        0.00       0.00
     3.00        2       6       2
     0       0       0       0
    30.00        0.00       0.00
     3.00        2       7       2
     0       0       0       0
    30.00        0.00       0.00
     3.00        2       8       2
     0       0       0       0
    30.00        0.00       0.00
     3.00        2       9       2
     0       0       0       0
    30.00        0.00       0.00
     3.00        2      10       2
     0       0       0       0
    30.00        0.00       0.00
     3.00        2      11       2
     0       0       0       0
    30.00        0.00       0.00
     3.00        2      12       2
     0       0       0       0
    30.00        0.00       0.00
```

Table 8-3 (continued)

```
$.4 Outflow Conditions
 $noutflows          0
 $qoutflow,ioutflow,joutflow,koutflow
 $4.Elevation Boundary Conditions
 $nelevation  kts              1           2
 $iewest,ieeast,jesouth,jenorth
 zmean,zamp,tmelag,tideper
 k,temp,saln,const,cbod,on,nh3,do(d),no3,op,po4,phyt for k=2,km-2
 20  20  2  12
 0.0 0.0  0.0  0.0
 2    30   0   0
 3    30   0   0
 4    30   0   0
 $5.Initialize Water Quality Profiles
 $ninitial               1
 k,temp,saln,const,cbod,on,nh3,do(d),no3,op,po4,phyt for k=2,km-2
 2    30   0   0
 3    30   0   0
 4    30   0   0
 $6.External Parameters
 Chezy, Wx,Wy,CSHE, TEQ, Rdecay, Lat.
   35.00     0.00     0.00    25.00    30.00    0.00    39.6
 $7.Output Profiles
 $nprofiles           0
 $ipwest,ipest,jpsouth,jpnorth
 $u-vel, v-vel, w-vel
 $nconstituents
 $I-const(1),I-const(2), I-const(3), etc
 $8.Output Surfaces
 $nsurfaces  $nconstituents          1           3
 $U-vel  v-vel  0.0000000E+00  0.0000000E+00
 $I-const(1), I-const(2), I-const(3), etc.
    3    4    5
 $9.Output Time Series
 $ntimser            2
 $nconst, iconst, jconst,kconst
    3    10   10    4
    4    10   10    2
 $10.Simulation time conditions
 $dtm  tmend    120.0000          120.0
 $tmeout  tmeserout   120      4.0
 $11.Internal Boundary Locations
 $nintbnd             1
 $ibwest,ibeast,jbsouth,jbnorth,ktop,kbottom
  10  10   2   6   2   6
 $12. Constituent Averages
 $nconarv            0
 $nconstarvs
 $13. Groundwater Inflow
 $ngrndwtr            0
```

9. GENERALIZED LONGITUDINAL LATERAL VERTICAL HYDRODYNAMIC TRANSPORT RELATIONSHIPS

The basic formulations and basis of the numerical computations in the Generalized Longitudinal Lateral Vertical Hydrodynamic and Transport model (GLLVHT) are presented in Edinger and Buchak (1980, 1985). Recent published peer reviewed applications include Edinger et al. (1994); Edinger and Buchak (1995); Edinger et al. (1998); Kolluru et al. (1998); Edinger and Kolluru (1998); Edinger and Kolluru (1999); Kolluru et al. (1999); Edinger and Kolluru (2000), and Buchak et al. (2001). The GLLVHT model has had more than 40 applications to different kinds of water body and water quality problems either in the INTROGLLVHT operating system or in the Generalized Environmental Modeling System for Surfacewaters (GEMSS), to which INTROGLLVHT is compared in Section 9.5.

9.1 Fundamentals of Momentum Balances

The formulation of the hydrodynamic relationships for free surface flows are as follows:

The horizontal x-momentum balance is taken as

$$DU/Dt = -1/\rho_o \; \partial P/\partial x \tag{9.1}$$

the horizontal y-momentum balance is taken as

$$DV/Dt = - \; 1/\rho_o \; \partial P/\partial y \tag{9.2}$$

and the vertical z-momentum balance is taken, with z positive downward, as

$$DW/Dt = g - 1/\rho_o \; \partial P/\partial z \tag{9.3}$$

where $D(\;)/Dt$ is the total transport derivative for the quantity $(\;)$ and for purposes of derivation is assumed to include the local change, advection, and dispersion of $(\;)$; U, V, and W are the velocity components in each of the x, y, and z directions; g is gravitational accleration; ρ_o is density; and P is pressure.

The numerical solution to the horizontal momentum balances requires the evaluation of the horizontal pressure gradients from the vertical pressure distribution as it depends on density and water surface elevation. Assuming that the vertical acclerations, DW/Dt, are small compared with gravitational acceleration (the hydrostatic approximation), the vertical momentum balance can be written as

$$1/\rho \; \partial P/\partial z = g. \tag{9.4}$$

The horizontal momentum relationships, Equations 9.1 and 9.2, require the horizontal pressure gradient as developed from the pressure. The pressure at any depth z is

$$P = g \int_{z'}^{z} \rho \, dz \qquad (9.5)$$

where z' is the free surface elevation. Differentiating Equation 9.5 and using Leibnitz's rule for differentiating an integral gives the horizontal pressure gradient in the x direction as

$$g/\rho o \partial P/\partial x = \partial z' /\partial x - g/\rho o \int_{z'}^{z} (\partial \rho /\partial x) \, dz. \qquad (9.6)$$

The first term on the right-hand side is the barotropic surface slope and the second term is the baroclinic slope.

9.2 Detailed Relationships

The hydrodynamic and transport relationships used in the GLLVHT are developed from the horizontal momentum balance, continuity, constituent transport, and the equation of state. The horizontal momentum balances for the horizontal velocity components, U and V in the x- and y-coordinate horizontal directions, with z taken positive downward, are

$$\partial U/\partial t = g \, \partial z'/\partial x - g/\rho \int_{z'}^{z} (\partial \rho /\partial x) \, \partial z - \partial UU/\partial x - \partial VU/\partial y - \partial WU/\partial z$$

$$-fV + \partial A_x(\partial U/\partial x)/\partial x + \partial A_y(\partial U/\partial y)/\partial y + \partial A_z(\partial U/\partial z)/\partial z \qquad (9.7)$$

$$\partial V/\partial t = g \partial z'/\partial y - g/\rho \int_{z'}^{z} (\partial \rho /\partial y) \partial z - \partial UV/\partial x - \partial VV/\partial y - \partial WV/\partial z$$

$$+fU + \partial A_x(\partial V/\partial x)/\partial x + \partial A_y(\partial V/\partial y)/\partial y + \partial A_z(\partial W/\partial z)/\partial z. \qquad (9.8)$$

Local continuity for the vertical velocity component W is

$$\partial W/\partial z = - \partial U/\partial x - \partial V/\partial y. \qquad (9.9)$$

Vertically integrated continuity for the surface elevation, z', integrated from the water surface elevation, z, to the bottom of the water column, z = h, is

$$\partial z'/\partial t = \int_{z}^{h} (\partial U/\partial x) \, dz + \int_{z}^{h} (\partial V/\partial y) \, dz. \qquad (9.10)$$

The constituent transport relationship for n number of constituents each with concentration Cn and a source/sink term of Hn is

$$\partial C_n/\partial t = -\partial U C_n/\partial x - \partial V C_n/\partial y - \partial W C_n/\partial z + \partial(D_x \partial C_n/\partial x)/\partial x$$

$$+ \partial(D_y \partial C_n/\partial y)/\partial y + \partial(D_z \partial C_n/\partial z)/\partial z + H_n. \tag{9.11}$$

And, the equation of state relating density, ρ, to constituents is

$$\rho = f(C_1, C_2, \ldots, C_n). \tag{9.12}$$

These relationships have six unknowns (U, V, W, z', ρ, C_n) in six equations, assuming that the momentum and constituent dispersion coefficients (A_x, A_y, A_z, D_x, D_y, D_z) can be evaluated from velocities and the density structure.

In the x and y momentum balances, the right-hand terms are successively the barotropic or water surface slope, the baroclinic or density gravity slope, the advection of momentum in each of the three coordinate directions, and the Coriolis acceleration. Next is the dispersion of momentum in each of the coordinate directions. The baroclinic slope is seen to be the vertical integral of the horizontal density gradient and becomes the major driving force for density-induced flows due to discharge buoyancy, and temperature and salinity stratification.

9.3 Computational Scheme

The hydrodynamic relationships are integrated numerically, implicitly forward in time, by evaluating the horizontal momentum balances as

$$\partial U/\partial t = g\partial z'/\partial x + F_x \tag{9.13}$$

$$\partial V/\partial t = g\partial z'/\partial y + F_y \tag{9.14}$$

where U, V, and z' are taken simultaneously forward in time and all the other terms are incorporated in the forcing functions F_x and F_y and are lagged in time. Equations 9.13 and 9.14 are substituted (either by cross-differentiation or algebraically from the finite difference forms) into vertically integrated continuity, Equation 9.10, to give the surface wave equation of

$$\partial^2 z'/\partial t^2 + g\partial(H\partial z'/\partial x)/\partial x + g\partial(H\partial z'/\partial y)/\partial y = \partial/\partial x \left(\int_{z'}^{h} F_x \partial z \right)$$

$$+ \partial/\partial y \left(\int_{z'}^{h} F_y \partial z \right) \tag{9.15}$$

where z' is the surface displacement and H is the total water column depth.

The computational steps in GLLVHT on each time step of integration are: (1) to evaluate F_x and F_y from U, V, W, ρ known from the previous time step; (2) to solve the surface wave equation for new z' for the spatial grid using a conjugate gradient solution method; (3) to solve for new U and V using Equations 9.13 and 9.14; (4) to solve for W using Equation 9.9; (5) to re-evaluate z' from W computed at the top of the top layer for precision; and (6) to solve the constituent relationship, Equation 9.11.

The semi-implicit integration procedure has the advantage that computational stability is not limited by the Courant condition that $\Delta x/\Delta t$, $\Delta y/\Delta t < (gh_m)^{1/2}$, where h_m is the maximum water depth, which can lead to inefficiently small time steps of integration. Because the solutions are semi-implicit (for example, explicit in the constituent transport and the time-lagged momentum terms) the stability is controlled by the Torrence condition ($U\Delta t/\Delta x$, $V\Delta t/\Delta y < 1$). Hence, the integration time step can be chosen to realistically represent the details of the boundary data, which is approximately 15 minutes for tides. Development of the computational steps in numerical form is presented in Section 9.4.

There are a number of auxiliary relationships that enter the computations. The vertical momentum dispersion coefficient and vertical shear is evaluated from a Von Karman relationship modified by the local Richardson number, Ri, (the ratio of vertical buoyant acceleration to vertical momentum transfer) as

$$A_z = kLm^2/2[(\partial U/\partial z)^2 + (\partial V/\partial z)^2]^{1/2}Exp(-1.5Ri) \qquad (9.16)$$

where k is the Von Karman constant, Lm is a mixing length that can be a function of depth or layer thickness, and Ri is the local Richardson number. The Richardson number function is from Leendertse and Liu (1975). The longitudinal and lateral dispersion coefficients are scaled to the dimensions of the grid cell using the dispersion relationships developed by Okubo (1971) of

$$Dx,Dy = 5.84 \ x10^{-4}(DXX,DYY)^{1.1} \qquad (9.17)$$

where Dx,Dy is the longitudinal or lateral dispersion coefficient in square meters per second and DXX,DYY is the longitudinal or lateral cell dimension in meters.

Wind surface stress enters the relationships for each of the coordinate directions as

$$\partial U/\partial z \ |_{z'} = CwWxAbs(Wx)/Az \qquad (9.18)$$

and

$$\partial V/\partial z \ |_{z'} = CwWyAbs(Wy)/Az \qquad (9.19)$$

where Wx and Wy are surface wind speed components in the x and y directions and Cw is the wind shear coefficient.

Bottom friction enters the computations through a Chezy friction relationship as

$$A_z\partial U/\partial z \ |_h = (g/C_h^2)Uabs(U) \qquad (9.20)$$

$$A_z\partial V/\partial z \ |_h = (g/C_h^2)Vabs(V) \qquad (9.21)$$

where C_h is the local Chezy friction coefficient and h is the bottom elevation at which bottom friction is evaluated.

The computational model is built to accept a large number of transport constituents and constituent relationships. The basic parameter obtained from the water quality model is the constituent flux, h(I,J,K,N). For example, for a simple decay constituent such as coliforms, $h(I,J,K,N) = -KR_4*C(I,J,K,3)$ for the decay of constituent 3 where KR_4 is the coliform decay constant. The constituent transport computation is explicit in time. It is developed so that transport coefficients can be computed once and used for all constituents during that time step at a given "n," "k" location. The solution time is not too sensitive to the number of constituents included in the model.

9.4 Formulation of the Generalized Longitudinal Lateral Vertical Hydrodynamic Transport Numerical Hydrodynamic Computation

The formulations of the momentum equations as actually set up are derived from the finite difference forms of the basic relationships. The space-staggered finite difference forms of the GLLVHT numerical computation used in INTROGLLVHT can be written from Equation 9.10, Equation 9.13, and Equation 9.14. Vertically integrated continuity in numerical form is

$$[Z(I,J)-Z0(I,J)]/\Delta t = 1/\Delta X \; \Sigma \; [U(I,J,K)-U(I-1,J,K)] \; \Delta Z$$

$$+ \; 1/\Delta Y \; \Sigma \; (V(I,J,K)-V(I,J-1,K)) \; \Delta Z \qquad (9.22)$$

where Z(I,J) is the water surface elevation at the new time step, Z0(I,J) is the water surface elevation at the old time step, U(I,J,K) is the x-velocity component on the new time step, V(I,J,K) is the y-velocity component on the new time step, ΔX is the cell length in the x direction, ΔY is the cell length in the y-direction, ΔZ is the layer thickness, and Δt is the iteration time step..

The X-momentum relationship is

$$[U(I,J,K)-U0(I,J,K)]/\Delta t = G \; [Z(I+1,J)-Z(I,J)]/\Delta X + Fx(I,J,K) \qquad (9.23)$$

where U0(I,J,K) is the x-velocity component at the old time step, G is the gravitational acceleration, and Fx(I,J,K) is the momentum forcing terms in the x direction.

The Y-momentum relationship is

$$[V(I,J,K)-V0(I,J-1,K)]/\Delta t = G[Z(I,J+1)-Z(I,J)]/\Delta Y + Fy(I,J,K) \qquad (9.24)$$

where V0(I,J,K) is the y-velocity component at old time step, and Fy(I,J,K) is the momentum forcing terms in the y direction.

1.1.1   The Surface Elevation Relationship

Substituting U(I,J,K) and V(I,J,K) from Equation 9.23 and Equation 9.24, respectively, into Equation 9.22 gives the surface elevation equation of

$$Z(I,J) = Z0(I,J) + G \, \Delta t^2/\Delta X^2 \, H(I,J) \, [Z(I+1,J)-2 \, Z(I,J)+Z(I-1,J)]$$

$$+ \, G \, \Delta t^2/\Delta Y^2 \, H(I,J)[Z(I,J+1)-2 \, Z(I,J)+Z(I,J-1)]$$

$$+\Delta t\Delta Z/\Delta X \, \Sigma \, [U0(I,J,K)-U0(I-1,J,K)]$$

$$+\Delta t\Delta Z/\Delta Y \, \Sigma[V0(I,J,K)-V0(I,J-1,K)]$$

$$+\Delta t^2\Delta Z/\Delta X \, \Sigma[Fx(I,J,K)-Fx(I-1,J,K)]$$

$$+\Delta t^2/\Delta Z/\Delta Y\Sigma[Fy(I,J,K)-Fy(I,J-1,K)] \tag{9.25}$$

where H(I,J) is the water column depth at the center of the cell.

Collecting the new time step elevations on the left-hand size, the surface elevation equation becomes

$$-G \, \Delta t^2/\Delta X^2 \, H(I,J) \, Z(I-1,J)- \, G \, \Delta t^2/\Delta Y^2 \, H(I,J) \, Z(I,J-1)$$

$$+ \, [1+2 \, G \, \Delta t^2/\Delta X^2 \, H(I,J)+ \, G \, \Delta t^2/\Delta Y^2 \, H(I,J)] \, Z(I,J)$$

$$-G \, \Delta t^2/\Delta X^2 \, H(I,J) \, Z(I+1,J)- \, G \, \Delta t^2/\Delta Y^2 \, H(I,J) \, Z(I,J+1) =$$

$$Z0(I,J) +\Delta t\Delta Z/\Delta X \, \Sigma \, [U0(I,J,K)-U0(I-1,J,K)] + \Delta t\Delta Z/\Delta Y \, \Sigma \, [V0(I,J,K)-V0(I,J-1,K)]$$

$$+\Delta t^2\Delta Z/\Delta X \, \Sigma \, [Fx(I,J,K)-Fx(I-1,J,K)]$$

$$+ \, \Delta t^2/\Delta Z/\Delta Y\Sigma \, [Fy(I,J,K)-Fy(I,J-1,K)] \tag{9.26}$$

On the left-hand side, if Z(I,J) is considered the center cell of a group of five cells, then Z(I-1,J), Z(I+1,J), Z(I,J-1), and Z(I,J+1) are the four adjacent cells. The new time step elevations on the left-hand side consist of five arrays in a matrix from which the new time step elevations can be computed at the new time step using a conjugate gradient computational algorithm. The main diagonal multiplier of Z(I,J) is always positive. The off-diagonal multipliers on the remaining Z terms are always negative. The matrix is perfectly symmetric, and the sum of the multipliers of the off-diagonal Zs is exactly equal to the multiplier of the main diagonal Z minus one, making it a perfectly weighted matrix with no bias in any direction.

On the right-hand side are the previous time step elevation, Z0(I,J), and the previous time step velocity components, U0(I,J,K) and V0(I,J,K), at their exact time level. The major assumption in the GLLVHT formulation is that the Fx(I,J,K) and Fy(I,J,K) are computed from old time step velocity components and densities. The Δt is limited by the accuracy of the solution and the Torrence condition as it applies to the advection of momentum.

1.1.2 Computation of the Velocity Components

Once $Z(I,J)$ is computed from Equation 9.26, the velocity components are computed from Equation 9.23 and Equation 9.24. Because Equation 9.26 is arrived at by substitution of the new time step velocity components, the $Z(I,J)$, $U(I,J,K)$, and $V(I,J,K)$ are essentially being computed simultaneously, except for lagging Fx and Fy in time. The solution technique is therefore considered to be semi-implicit in time for surface elevations and velocity components.

The $W(I,J,K)$ velocity component is found by integrating the internal continuity relationship, Equation 9.9, from the bottom upward to give

$$W(I,J,K-1) = W(I,J,K) - 1/\Delta X \; \Sigma[U(I,J,K) - U(I-1,J,K)] \; \Delta Z$$

$$- 1/\Delta Y \; \Sigma[V(I,J,K) - V(I,J-1,K)] \; \Delta Z \qquad (9.27)$$

which starts from $W[I,J,K0(I,J)] = 0$ at $K = k0(I,J)$ to the water surface at $k = kt$.

Integration to the surface gives $W(I,J,Kt-1)$, which is used to re-evaluate $Z(I,J)$ for accuracy as

$$Z(I,J) = Z0(I,J) + W(I,J,Kt-1)\Delta t. \qquad (9.28)$$

Equation 9.28 ensures that vertically integrated continuity is completely satisfied and that there is an exact water balance in the variable thickness top layer.

9.5 Numerical Formulation of the Generalized Longitudinal Lateral Vertical Hydrodynamic Transport Relationship

Once the velocity components are known to be semi-implicit in time computation and the source/sink terms have been evaluated for each constituent, the constituent distribution is computed explicitly forward in time from Equation 9.11. The constituent computation in INTROGLLVHT uses the upwind differencing approximation on the advection terms that can lead to numerical dispersion in regions of sharp concentration gradients.

The transport relationship uses the following upwind operators and logic:

$$U1 = 1 \text{ when } U(I,J,K) > 0 \qquad (9.29a)$$

$$U2 = 1 \text{ when } U(I-1,J,K) > 0 \qquad (9.29b)$$

$$V1 = 1 \text{ when } V(I,J,K) > 0 \qquad (9.29c)$$

$$V2 = 1 \text{ when } V(I,J-1,K) > 0 \tag{9.29d}$$

$$W1 = 1 \text{ when } W(I,J,K) > 0 \tag{9.29e}$$

$$W2 = 1 \text{ when } W(I,J,K-1) > 0 \tag{9.29f}$$

When the stated condition of a positive velocity component is not satisfied, then the upwind operator is zero. These are used in the advective terms which are formulated for example as

$$\partial(UC)/\partial x = [U(I,J,K)*(U1*(C(I,J,K,N))+(1-U1)*C(I+1,J,K,N)]$$

$$- U(I-1,J,K)*[U2*(C(I-1,J,K,N)+(1-U2)*C(I,J,K,N))]/\Delta x. \tag{9.30}$$

The constituent concentrations $C(I,J,K,N)$ for constituent N are centered in each cell. The velocities are at the cell interfaces. The first term on the right-hand side of Equation 9.30 states that when $U(I,J,K)$ is positive, it removes the constituent from the cell, and when negative moves the constituent in from the adjoining I+1 cell. Similar advective terms can be written in the Y and Z coordinate directions.

All of the terms in the constituent transport relationship (Equation 9.11) are expanded and rearranged into the following form:

$$C(I,J,K,N) = f0*C0(I,J,K,N)+f1*C0(I+1,J,K,N)+f2*C0(I-1,J,K,N)+f3*C0(I,J+1,K,N)$$

$$+f4*C0(I,J-1,K,N)+f5*C0(I,J,K,N)+f6*C0(I,J,K-1,N)+h(I,J,K,N)*\Delta t \tag{9.31}$$

where $C(I,J,K,N)$ is the constituent concentration computed at the new time step from the $C0(I,J,K,N)$ of the old time step, and $h(I,J,K,N)$ is the source/sink term for the particular water quality model being used. It is related to rate relationships in the different water quality models as $h(I,J,K,N) = DC/Dt$. The new $C(I,J,K,N)$ is computed from the transport contribution for each of the adjacent cells.

The computation proceeds one cell at a time as the functions in Equation 9.31 are computed for that cell. The computation uses these functions to compute the new concentration of each constituent in that cell before proceeding to the next cell. The functions in Equation 9.31 resulting from the collection of terms of the expansion of Equation 9.11 to finite difference form are

$$f0 = \{1-[U1*U(I,J,K)-(1-U1)*U(I-1,J,K)]\ \Delta t/\Delta x-[V1*V(I,J,K)-(1-V1)*V(I,J-1,K)]\ \Delta t/\Delta y$$

$$- [W1*W(I,J,K)-(1-W1)*W(I,J,K-1)]\ \Delta t/\Delta z - [DX(I,J,K)+DX(I-1,J,K)]\ \Delta t/\Delta x^2$$

$$-[DY(I,J,K)+DY(I,J-1,K)]\ \Delta t/\Delta y^2 - [DZ(I,J,K)+DZ(I,J,K-1)]\ \Delta t/\Delta z^2 \} \tag{9.32a}$$

where the $DX(I,J,K)$, $DY(I,J,K)$, and $DZ(I,J,K)$ are the local turbulent dispersion coefficients across each of the cell faces.

Next,

$$f1 = -U(I,J,K)*(1-U1) \, \Delta t/\Delta x + DX(I,J,K) \, \Delta t/\Delta x^2 \qquad (9.32b)$$

$$f2 = U(I-1,J,K)*U1 \, \Delta t/\Delta x + DX(I-1,J,K) \, \Delta t/\Delta x^2 \qquad (9.32c)$$

$$f3 = -V(I,J,K)*(1-V1) \, \Delta t/\Delta y + DY(I,J,K) \, \Delta t/\Delta y^2 \qquad (9.32d)$$

$$f4 = V(I,J-1,K)*V1 \, \Delta t/\Delta y + DY(I,J,K) \, \Delta t/\Delta y^2 \qquad (9.32e)$$

$$f5 = -W(I,J,K)*(1-W1) \, \Delta t/\Delta z + DZ(I,J,K) \, \Delta t/\Delta z^2 \qquad (9.32f)$$

$$f6 = W(I,J,K-1)*W1 \, \Delta t/\Delta z + DZ(I,J,K-1) \, \Delta t/\Delta z^2 \qquad (9.32g)$$

The computation of the transport of all constituents in a given cell using the functions in Equation 9.32a to Equation 9.32g results in a very efficient, not lengthy simulation for a large number of constituents.

The constituent transport computation is explicit in time and requires that f0 always be positive. This requires in Equation 9.32a that the sum of the advection and dispersion terms, as scaled to the time step and their gradient distances, be less than unity. This condition, known as the Torrence condition, controls the integration time step of the computation.

9.6 INTROGLLVHT Model Limitations

The limitations on the INTROGLLVHT modeling system can be identified by comparison to a more highly developed computational system. INTROGLLVHT is compared to the GEMSS computational framework described in detail on the model summary site www.jeeai.com. The limitations of the INTROGLLVHT modeling in comparison with GEMSS are listed in Table 9-1.

Table 9-1. Comparison of introductory GLLVHT and complete Generalized Environmental Modeling System for Surfacewaters Modeling System (GEMSS).

Property	Introgllvht	GEMSS Modeling System	Advantages of GEMSS System
1) IM, JM, KM	50, 50, 30	Machine limits	Very large grids
2) ΔX, ΔY, ΔZ	Constant in each direction	Variable from cell to cell; curvilinear; variable thickness ΔZ at each elevation.	Fit shorelines precisely, provide more refined grid detail where needed; each cell has its own orientation of winds and for accurate orientation of winds and for mapping; can map to GIS
3) Layer-cell addition subtraction	No	Full capability	Represent large tidal fluctuations in thin layers; wet and dry tidal flats and beaches due to tidal fluctuations or changes in reservoir levels
4) Interior boundaries	Yes	Yes	Representation of interior structures such as breakwaters, marinas, underflow/overflow curtain walls
5) Vertical momentum	Hydrostatic approximation	Included; relaxes hydrostatic approximations (a)	Important for drawdown at outflow structures, mixing devices, and accurate representation of water surfaces in regions of large horizontal velocity changes
6) Location of inflows	Fixed specified I, J, and K location	Inflow location determined by inflow density relative to receiving water body density gradient; moves I, J, and K location with the addition and subtraction of layers and cells	Realistic density inflow conditions
7) Discharge momentum	None included	All three directions	Used for proper representation of high velocity discharges
8) Coriolis acceleration	Included; constant latitude	Variable with latitude	Can do large water bodies accurately over large variations in latitude
9) Transport scheme	Upwind differencing	Quickest, ultimate	Better prediction of constituent profiles in regions of sharp changes
10) Vertical dispersion and shear	Von Karmon relation	Von Karmon or higher-order schemes	Better description of turbulence in regions of rapid changes in bathymetry and around structures; also at density interfaces
11) Horizontal dispersion	Scaled to ΔX, ΔY	Scaled to ΔX, ΔY or from vorticity (b)	Vorticity relationship completely dynamic and symmetric around cell; requires field velocity time series data or APC profiling data for calibration.
12) Wind speed	Constant and uniform over the grid	Variable through time and across grid	Realistic representation of wind events on a water body
13) Surface heat exchange	CSHE, Teq	Time-varying term by term heat budget	Accurate representation of diurnal variations in heat exchange from hourly or more frequently available data
14) Highest level eutrophication water quality model	WQ3DP particulate-based water quality model; parameters set up in tabular format by	WQ3DP and WQ3DCB coupled with sediment exchange model; parameters in both evaluated through a GUI	More realistic representation of processes taking place; WQ3DCB now includes dinoflagellates with vertical light dependent vertical migration; GUI interface for setting parameters based on flowchart of the model; WQ3DCB can be linked to the sediment diagenisis model

Property	IntrogIlvht	GEMSS Modeling System	Advantages of GEMSS System
15) Sediment exchange (diagenisis) model	hand; Empirical relationships or measured values	WQ3DCB carbon-based model coupled to diagenisis model	Coupling to a diagenisis model gives a complete water quality modeling system
16) Separate contribution of different sources	Can handle multiple discharges, but cannot separate out contribution	Can separate out contribution of each discharge	A very useful tool when having to evaluate individual discharge contributions to a particular water quality limit as is required for example in TMDL studies
17) Other supported routines and processes	Simplified sediment scour/deposition routine	Sediment transport; spill model; toxics model; intake entrainment model	Additional routines can be included in a modular fashion and run directly in GEMSS on a real-time basis
18) Grid development	By hand	Gridgen routine from digitized maps	Rapid grid development and changes
19) Input data	Constant values	Time-varying input data for all variables	Simulation of real-time conditions over seasons and years depending on limitations of input data base.
20) Outputs	Tablular results; time series can be plotted in spread sheets; vector and constituent contour plotting requires additional software such as TecPlot© or MatLab©	Complete integrated post processor including tabular and three-dimensional displays of constituent distributions and velocity vectors in each coordinate plane and as profiles; grid stores field data	Display time-varying results; animated displays; mappable onto GIS displays; convenient direct comparison of simulation results and field data within computational grid
21) Computational model setup	Bathymetry and input data tables set up manually in specific limited format	Graphical user interfaces used for entering input data, linking time varying data files, and locating inflows and outflows from grid map	Handles complex arrangements of model setup with internal help files
22) Specification of spatial file output frequencies	Constant for all specified profile slices and surfaces	Independent output start-stop times and frequencies can be specified separately for each output	Manageable outputs at time of available field data for verification
23) Restart capabilities	Must begin simulations each time from time zero; runs on a simulation time clock	Stores last step of velocity field water surface elevations and constituent distributions; runs on a real-time clock	Easy to restart simulations for new set of external parameters inputs; real-time clock provides a convenient basis for lining up all of the time varying input files
24) Verification against actual field data	Only qualitatively; very few real water body problems are for steady-state or	Almost every project to which it has been applied	Main calibration history of GEMSS has been to have accurate bathymetry and accurate input data for doing real-time simulations; verification has been performed against detailed time series velocity records and detailed instantaneous water quality profiles

Property	Introgllvht	GEMSS Modeling System	Advantages of GEMSS System
	stationary-state conditions		
25) Direct comparison of model results and field data	None; must be performed manually	Fully specified through graphical user interfaces	Rapid comparison of model results and field data
26) User-defined interfaces	None	Yes	Allows user to write routines particular to their analysis; for example, routines can be written for evaluating the rate parameters as a function of temperature, location in the water column, and nonlinear interactions

(a) Edinger and Kolluru 1999.

(b) Smagorinsky, J 1963

10. FIRST-ORDER DECAY AND SEDIMENT RELATIONSHIPS

The first-order decay can be used to approximate the fate of many substances in water including chlorine, coliforms, radio nuclides, and many forms of organic compounds. It is also used in the INTROGLLVHT modeling system to compute the residence time or flushing time distribution throughout a water body.

10.1 First-Order Decay

The first-order decay of a substance is formulated in the numerical model from

$$DC/Dt = -Rdecay\ C \tag{10.1}$$

where DC/Dt represents the local change in storage and all of the advection and dispersion terms.

10.2 Decay Rates

Typical ranges of first-order decay rates for a few different constituents are presented in Table 10-1. The chlorine decay rates are taken from Davis and Coughlan (1981) and the remaining decay rates are from Howard et al. (1991).

Almost all of the organic constituents have decay products that may be more toxic than the parent product. An example is the decay product vinyl chloride that may result from the decay of chlorine. Also, many products may now be banned from discharges, particularly herbicides.

10.3 Analytical Solution

The first-order decay relationship has a simple analytical solution over time of

$$C = Co\ Exp(-Rdecay\ t) \tag{10.2}$$

where Co is the concentration at $t = 0$. The time can also be taken as the time of travel down a simple channel as $t = X/U$.

10.4 Half-Life from Kinetic Coefficients

The half-life for a given Rdecay can be computed from Equation 10.2 as the time at which $t = t_{1/2}$ for $C/Co = 0.50$. Solving Equation 10.2 for this condition gives the following:

$$t_{1/2} = 0.693/Rdecay. \tag{10.3}$$

The half-life for different kinetic rates from the different water quality models can be compared with the residence times in different locations within a water body to see where that portion of the process is most likely to go to completion.

10.5   Computation of Residence Time

The computation of residence time, or flushing time, is included in INTROGLLVHT. The input file setup required for computing residence time is presented in Chapter 4, Section 4.2. It is defined as the time required for the initial virtual dye concentration to reduce by one half.

The flushing time is defined as a half-life as

$$Exp(-k*t_{1/2}) = 0.50 \qquad\qquad (10.4)$$

where the decay rate, k, applies only to an individual model box.

At the end of the simulation, the dye concentration is assumed to be

$$C/C_o = Exp(-k*tmend) \qquad\qquad (10.5)$$

where Co is the dye concentration at which the dye is initialized in the water body.

Eliminating k between (10.4) and (10.5) gives the following:

$$t_{1/2} = Ln(2)/Ln(C_o/C)*tmend/24, \text{ days} \qquad\qquad (10.6)$$

where C is the dye concentration in the box.

Table 10-1. Decay rates for selected constituents.

Constituent	Rate range, per day
Aldrin	0.033
Coliforms	0.50 – 2.0
Chloroform	0.025
Chlordane	0.002-0.004
Chlorine	0.027 (18 C)-0.042 (33 C)
Dieldrin	0.0039
Flourene	0.022
Heptachlor	0.129
Malathion	0.028
Methanol	0.099
Trichloroethylene	0.0039
Vinyl chloride	0.0247

11. SURFACE HEAT EXCHANGE RELATIONSHIPS

The surface heat exchange in the INTROGLLVHT model is computed using the coefficient of surface heat exchange (CSHE) and the equilibrium temperature relationship. Both the full term-by-term surface heat exchange relationships and the CSHE/Teq relationships are developed in Edinger et al. (1974). A brief derivation of the latter is presented here.

11.1   Approximation to Surface Heat Exchange Formulation

The net rate of surface heat exchange is defined as follows:

$$Hn = Hs + Ha - Hbr - He - Hc \qquad (11.1)$$

where:

Hs = short-wave solar radiation, W/m^2

Ha = long-wave atmospheric radiation, W/m^2

Hbr = back radiation from the water surface, W/m^2

He = evaporation from the water surface, W/m^2

Hc = conduction between water and air, W/m^2

Over the long term (5 to 10 days), $Ta \cong Tw$, where

Ta = air temperature, C

Tw = water temperature, C,

therefore, $Ha \cong Hbr$, and Hc is a very small part of the net heat exchange. In the long term, therefore, the net rate of surface heat exchange reduces to

$$Hn = Hs - He. \qquad (11.2)$$

Let

$$He = \beta f(w)(Tw - Td) \qquad (11.3)$$

where

β = vapor pressure slope, mmHg/degC

Td = dew point temperature, C

$f(w)$ = evaporative wind speed function, $W/m^2/mmHg$

171

Substituting Equation (11.3) into (11.2) gives the net rate of heat exchange as follows:

$$Hn = Hs + \beta f(w)Td - \beta f(w)Tw. \qquad (11.4)$$

Letting $CSHE(W/m^2/degC)$ be defined as follows:

$$CSHE = dHn/dTd \qquad (11.5)$$

gives

$$CSHE = \beta f(w) \qquad (11.6)$$

and

$$Hn = Hs + CSHE(Td - Tw). \qquad (11.7)$$

The equilibrium temperature Teq is defined as follows:

Teq = Tw when Hn = 0,

giving

$$Teq = Hs/CSHE + Td. \qquad (11.8)$$

Equation (11.8) is known as the Brady approximation to the equilibrium temperature of surface heat exchange.

Solving Equation (11.8) for Hs and substituting into Equation (11.7) gives the net rate of surface heat exchange as follows:

$$Hn = -CSHE(Tw - Teq). \qquad (11.9)$$

11.2   Excess Temperature

The increase in water temperature due to a heat source, θ, is called the excess temperature. The net rate of surface heat exchange with the heat source then becomes:

$$Hn' = -CSHE(Tw + \theta - Teq). \qquad (11.10)$$

Subtracting Equation (11.9) from (11.10) gives the net rate of surface heat exchange due to the heat source alone as follows:

$$Hn' - Hn = -CSHE \, \theta. \qquad (11.11)$$

The excess temperature can be evaluated using the TSC model by setting Teq equal to 0 and specifying only the temperature rise of the discharge. No inflow initial or boundary temperatures should be specified to carry out the excess temperature computation.

11.3   Short-Wave Solar Radiation

Values of clear sky short-wave solar radiation are given in Table 11-1. The actual short-wave solar radiation reaching the water surface is

$$Hs = (1-0.0071Cld^2)Hsc \tag{11.12}$$

where

Hs = short-wave solar radiation

Cld = cloud cover in tenths (ranging from 1 for low cloud cover to 10 for complete cloud cover)

Hsc = clear sky short-wave solar radiation

The cloud cover is available through the National Oceanic and Atmospheric Administration climatologic data sites.

11.4   Coefficient of Surface Heat Exchange and Equilibrium Temperature

The coefficient of surface heat exchange is a function of the wind speed and the average of the water surface temperature and dew point temperature as $Tm = (Ts+Td)/2$. Table 11-2 gives the values of the coefficient of surface heat exchange in watt/m^{2}/C for a range of Tm and wind speeds.

The Brady approximation, Equation 11.8, shows that the computation of Teq and CSHE from the meteorologic data of Hs, Td, and W is an iterative process that can be tedious to do as a hand computation. The routine given in the INTROGLLVHT folder named Teq and Cshe Computation.exe performs this computation.

The routine asks for clear sky radiation, which is available from Table 11-1, the cloud cover in tenths, the dew point temperature, and the wind speed. It also asks for the average depth of the water body to estimate how long it takes the water body to reach equilibrium. The routine prints out to the screen Teq, CSHE, and the time of approach to equilibrium. No file is necessary to store these results.

11.5   Response Temperature

A useful quantity for estimating the response time for a water body to approach equilibrium from initial temperatures is the response temperature. The response temperature is the temperature a water body would reach due to surface heat exchange alone. For a mixed water column, it can be defined as follows:

$$\rho CpD \; dTr/dt = -CSHE*(Tr-Teq) \tag{11.13}$$

where:

ρ = density of water (1000 kg/m^3)

Cp = specific heat of water (4186 joules/kg/degC)

D = depth of the water column, m

CSHE = coefficient of surface heat exchange, (15-35 Watt/m^2/C)

The solution to Equation (11.13) for a water body beginning at an initial temperature of Ti is over time:

$$Tr(t) = TiExp(-Kt)+Teq[1-Exp(-Kt)] \tag{11.14}$$

where:

$$K = CSHE/(\rho CpD). \tag{11.15}$$

Equation (11.14) can be applied to a simple stream case where Ti is the mixed temperature at the head of the stream and Teq is the temperature the stream would approach as a function of travel time.

The relationship for excess temperature would be as follows:

$$d\theta/dt = -K\theta \tag{11.16}$$

and the decay of a fully mixed excess temperature in the downstream direction would be

$$\theta(t) = \theta_0 \; Exp(-Kt) \tag{11.17}$$

where θ_o is the fully mixed excess temperature computed as Qp θ_p/Qr and θ_p is the temperature rise across the facility, Qp is the facility low rate, and Qr is the river flow rate.

11.6   Plant Heat Rejection, Pumping Rate, and Temperature Rise

The relationship between a steam electric power plant heat rejection rates, pumping rates, and temperature rise can be illustrated for the cooling reservoir examined in Chapter 7. A fairly heavily loaded closed loop cooling lake is typically loaded at 1 Megawatt electric (Mwe) per acre. For the example reservoir the loading could then be as high as 2400 Mwe. Typically, a steam electric plant is approximately 30% efficient, and at this efficiency a 2400 Mwe plant would have a total heat rejection of 5600 Mwth (Megawatt thermal). For a fossil plant, approximately 80% of this heat is rejected to the cooling water, and the rest goes up the stacks. The heat rejection to the cooling water is, therefore, 4480 Mwth.

For modeling purposes it is necessary to specify the pumping rate of the plant and the temperature rise across the plant. These are directly related to the plant heat rejection. The temperature rise across the plant, or the condenser temperature rise, is typically 10° C (18° F). The plant heat rejection, flow rate, and condenser temperature rise are related as

$$Hp = \rho \ Cp \ Qp \ \Delta Tp \qquad\qquad (11.18)$$

where:

Hp = plant heat rejection in Mwth x 10^6 to give Wth

ρ = density of water, 1000 kg/m^3

Cp = specific heat of water, 4186 joules/kg/degC (1 joule/sec = 1 Watt)

Qp = plant pumping rate, m^3/s

ΔTp = condenser temperature rise, degrees C

For ΔTp = 10 $^\circ$C and Hp = 4480 Mwth, the plant pumping rate is 107 m^3/s. A check which can be made prior to simulation is to estimate the average temperature rise at the reservoir surface to dissipate the waste heat. The relationship between heat rejection, reservoir surface area, and temperature rise on a completely mixed reservoir is as follows:

$$Hp = Cshe \ \Delta Ts \ As \qquad\qquad (11.19)$$

where:

$Cshe$ = coefficient of surface heat exchange, watt/m^2/degC

ΔTs = increase in surface temperature, degrees C

As = reservoir surface area, m^2

For a $Cshe$ of 35 watt/m^2/degC and a reservoir surface area of 9.8 x 10^6 m^2, the estimated average surface temperature increase is 13° C. The surface temperatures on the recirculating reservoir can be compared to this to see if it is more efficient than a completely mixed reservoir.

Buchak et al. (2001) give further details about evaluating cooling water discharge sizes, the use of cooling lakes, and real-time modeling and data comparisons.

Table 11-1. Daily average clear sky solar radiation, W m^{-2*}.

Latitude	Jan	Feb	Mar	Apr	May	Jun	Jul	Aug	Sep	Oct	Nov	Dec
47	55.0	101.6	170.7	244.2	303.2	332.2	323.9	280.3	212.7	138.8	77.6	45.0
46	61.5	108.7	177.3	249.6	306.9	334.2	324.5	280.3	213.1	140.3	80.7	49.9
45	65.9	111.6	178.9	250.5	307.6	335.6	327.1	284.4	218.5	146.5	87.2	55.8
44	70.9	116.0	182.5	253.1	309.6	337.3	329.0	286.8	221.8	150.7	92.0	61.0
43	76.4	120.7	186.2	255.8	311.4	338.7	330.5	289.0	224.9	154.9	97.2	66.6
42	81.7	125.5	190.0	258.5	313.2	340.0	331.9	291.0	227.8	158.9	102.1	72.0
41	86.8	129.6	193.0	260.4	311.5	341.1	333.4	293.3	231.3	163.4	107.3	77.5
40	92.5	134.9	197.3	263.6	316.4	342.2	334.2	294.5	233.4	166.8	111.9	82.9
39	97.8	139.3	200.6	265.7	317.8	343.3	335.6	296.8	236.8	171.3	117.3	88.7
38	103.3	144.1	204.2	268.1	319.1	344.0	336.3	298.1	239.3	175.1	122.1	94.2
37	108.3	147.7	206.5	269.2	319.6	344.6	337.8	300.9	243.5	180.5	128.2	100.1
36	113.3	151.5	208.8	270.3	320.0	345.1	339.0	303.3	247.3	185.5	134.0	106.0
35	119.4	157.6	213.9	273.8	321.7	345.3	338.3	302.6	247.5	187.2	137.4	111.1
34	124.3	161.0	215.8	274.5	321.7	345.3	339.1	304.7	251.2	192.2	143.2	116.9
33	130.0	166.2	219.8	276.9	322.7	345.4	339.0	305.1	252.7	195.2	147.6	122.3
32	135.4	171.0	223.6	279.5	324.3	346.3	339.8	306.6	255.1	198.9	152.4	127.7
31	139.8	172.4	222.5	277.0	321.8	345.3	341.4	311.1	262.2	207.5	161.1	135.1
30	145.9	179.1	228.4	281.2	323.7	344.8	339.2	308.3	260.0	207.0	162.9	139.2
29	152.3	184.9	233.0	284.1	324.9	344.9	338.8	308.3	261.2	209.9	167.5	145.1
28	156.8	187.3	233.5	283.2	323.8	344.5	340.2	311.7	266.7	216.7	174.7	151.6
27	162.3	192.0	236.7	284.9	324.1	344.1	339.7	312.0	268.3	219.9	179.3	157.1
26	166.6	194.0	236.6	283.3	322.0	342.8	340.1	311.8	273.3	226.4	186.3	163.5

*Based on equations shown in Environmental Protection Agency, 1971. Values are for the middle of the indicated month. Latitudes range from southern Texas (26°N) to Washington State (47°N).

Table 11-2. Coefficient of surface heat exchange in Watt/m^2/°C as a function of wind speed and mean temperature Tm*.

		Wind Speed m/s					
		1.5	2	2.5	3	3.5	4
	10	11	11	12	13	14	15
	15	13	14	15	16	17	18
Tm, C	20	16	17	18	20	21	23
	25	20	21	22	24	26	29
	30	24	25	27	30	32	36
	35	29	30	33	36	39	43

*Computed from relationships given in Edinger et al. (1974).

12. DISSOLVED OXYGEN DEPRESSION RELATIONSHIPS

The DOD model determines the uptake of dissolved oxygen due to the decay of ammonium (NH_4) and carbonaceous biochemical oxygen demand (CBOD). It also includes the mineralization of organic nitrogen (ON) to NH_4 and the latter to nitrate (NO_3). In this chapter, the fundamental rate relationships used in the DOD model are presented along with an analytical solution for a simple river channel case.

12.1 The Dissolved Oxygen Depression Processes

The DOD model includes the decay of ON to NH_4, the decay of NH_4 to NO_3 with the uptake of dissolved oxygen, and the decay of CBOD to carbon dioxide (CO_2) with the uptake of dissolved oxygen.

1.1.3 Decay of Organic Nitrogen

The equation for the decay of ON (NO) to ammonia–ammonium ($NH_3 + NH_4$) is as follows:

$$\frac{\partial No}{\partial t} = - KonNo \tag{12.1}$$

where Kon is the organic nitrogen decay rate.

1.1.4 Decay of Biochemical Oxygen Demand

The equation for the decay of BOD (L) is

$$\frac{\partial L}{\partial t} = - Kbod\,L \tag{12.2}$$

where Kbod is the decay rate of BOD.

1.1.5 Decay of Ammonia–Ammonium

The equation for the decay of ammonia–ammonium NH_3+NH_4 (Nh) to NO_3 is

$$\frac{\partial Nh}{\partial t} = - KnhNh + KonNo \tag{12.3}$$

where Knh is the decay rate of NH_4.

1.1.6   Uptake of Dissolved Oxygen by Ammonium

The NH_4 takes up dissolved oxygen as it decays to NO_3 as follows:

$$NH_4 + 2\,O_2 \Rightarrow NO_3 + H_2O + 2\,H \qquad (12.4)$$

(14 gm) (64 gm),

which shows that every 14 gm of NH_4 takes up 64 gm of oxygen (O_2) when decaying to NO_3. The ratio of dissolved oxygen loss to the rate of decay of ammonia is, therefore, $\alpha_N = 4.57$ theoretically, although it can vary between 4.33 and 4.57 due to the uptake of NH_4 in the cellular structure of bacteria carrying out the reaction.

1.1.7   Dissolved Oxygen Depression

The rate at which the dissolved oxygen is depressed is

$$\frac{\partial D}{\partial t} = KbodL + \alpha_N KnhNnh - RsD / dz \qquad (12.5)$$

where D is the dissolved oxygen depression from background values due to an inflow of ON, NH_4, and BOD. The surface reaeration rate of dissolved oxygen is Rs, and it applies only to the top layer of the model, which has a thickness of dz. In multilayered water bodies, Rs is usually taken as a function of wind speed using the Mackay(1980) formulation, which is

$$Rw(m/d) = 0.055*Wad^{1.5} \qquad (12.6)$$

for the surface reaeration wind speed, Wad, in m/s. The Wad does not necessarily need to be the same as the Wx and Wy used for surface wind shear.

12.2   Analytical Solution of the Dissolved Oxygen Depression Relationship

For the discharge of NH_4 and BOD into a fully mixed stream, there is an analytical solution to the dissolved oxygen depression relationship as a function of travel time along the stream. This is an extension of the classical Streeter-Phelps solution to the oxygen sag problem.

The solution of Equation (12.2) for BOD is as follows:

$$L = -L_0\, e^{-K_L t} \qquad (12.7)$$

where L_0 is the mixed concentration of the discharge of BOD in the stream and K_L is the BOD decay rate.

The solution of Equation (12.3) is

$$Nnh = -N_0 e^{-K_N t} \tag{12.8}$$

where N_0 is the mixed concentration of the discharge of NH_4 in the stream and K_N is the ammonium decay rate. Substituting Equation (12.6) and (12.7) into Equation (12.5) and solving the resulting equation gives the dissolved oxygen depression as a function of travel time along the fully mixed stream as follows:

$$D = \frac{K_L L_0}{r - K_L} (e^{-K_L t} - e^{-rt}) + \frac{\alpha_N K_N N_0}{r - K_N} (e^{-K_N t} - e^{-rt}) . \tag{12.9}$$

12.3   Reaeration Formulas for Streams

There are many formulations for determining the reaeration rate, r, from stream velocity and stream depth (EPA 1985). Churchill et al. (1962) provide a formula based on extensive stream data that gives

$$r = 5.03 \ U^{0.969}/H^{1.673} \tag{12.10}$$

for r in units of per day. The U is the stream velocity in m/s determined from the flow divided by the cross section, and H is the stream depth in meters determined from the stream cross section divided by the stream surface width. The Churchill formula is based on data for stream velocities ranging from 0.50 to 1.5 m/s, and stream depths ranging from 0.5 to 3.5 m.

The INTROGLLVHT model uses the Mackay (1980) surface wind speed formula as indicated previously. For simple vertically mixed streams it is necessary to use an *equivalent wind speed* that gives the same results as the Churchill et al. (1962) formulation.

Letting Rw(m/d) = H*r, then an *equivalent wind speed* is

$$Wad = 19.64*U^{0.54}/H^{0.44}. \tag{12.11}$$

The equivalent wind speed allows one to apply the three-dimensional model to simple streams where reaeration is a function of velocity and depth.

13. THE WATER QUALITY DISSOLVED PARTICULATE EUTROPHICATION MODEL

The particulate-based WQDP eutrophication model is an extension of EUTRO5 incorporated in the Environmental Protection Agency's WASP5 (Ambrose et al. 1993). The WQDPM is similar to the Qual2E model (Brown and Barnwell 1987) that incorporates the particulate forms of particulate carbonaceous biochemical oxygen demand (CBOD_p), ON_p, and OP_p. The Qual2E model applies only to one-dimensional streams, whereas the WQDPM is three dimensional. In addition, the WQDPM incorporates grazing zooplankton whose excretions make up a large part of the particulate constituents. Other water quality models also include the particulate forms of the nutrient constituents, including the CE-QUAL-W2 longitudinal–vertical reservoir and estuary model (Buchak and Edinger 1984; Cole and Buchak 1995) and the CE-QUAL-ICM model (Cerco and Cole 1993). WQDPM is coupled to the GLLVHT hydrodynamic and transport model.

13.1   The Water Quality Dissolved Particulate Model Nutrient Cycles and Processes

The flow chart for the WQDPM is shown in Figure 13-1. The chart shows each of the 11 constituents incorporated in the WQDPM and the cycling of nutrients between the phytoplankton and the constituents. The dissolved carbonaceous biochemical oxygen demand (CBOD_d) and CBOD_p take up dissolved oxygen (DO) as they decay to CO_2, but are resupplied by the death of and excretion from the phytoplankton, and by wastes from the grazing zooplankton. The particulate form settles. Both organic nitrogen (ON) and organic phosphorous (OP) have dissolved and particulate forms that decay and take up DO, and are resupplied by death and excretion for the phytoplankton and wastes from the grazing zooplankton. The decaying OP produces phosphate (PO_4) that is taken up by the phytoplankton. The decaying ON produces ammonium (NH_4), which in turn decays to nitrate (NO_3), taking up DO. The phytoplankton use both NH_4 and NO_3; they prefer NH_4 until it reaches very low concentrations, and then they use NO_3.

The DO is taken up by the phytoplankton by respiration, and it is produced by the phytoplankton by photosynthesis. The DO is supplied by reaeration at the water surface, and taken up by sediment oxygen demand (SOD) in the bottom. The phytoplankton, CBOD_p, ON_p, and OP_p settle to the bottom. The sediment releases NH_4 and PO_4 back into the cycle.

13.2   The Water Quality Dissolved Particulate Model Constituent Relationships

The constituents included in the WQDPM are listed in Table 13-1, which gives the reference number used in INTROGLLVHT. The constituent symbol as used for each constituent in the constituent relationships is given in Table 13-2. The constituent symbols used in Table 13-2 are extensions of the EUTRO5 notation. In Table 13-2, the DCn/Dt is the total transport derivative and is the sum of the local change in constituent concentration, the advection of constituent, and the turbulent dispersion of constituent. Each constituent rate relationship is discussed separately. The individual parameters in each relationship are listed alphabetically in Table 13-3 and discussed in Section 13.3.

13.2.1 Ammonia and Ammonium Nitrogen

The NH_4 relationship is represented in Equation 13.1.1 in Table 13-2. Within the cycle it is supplied by dying phytoplankton and the mineralization of dissolved and particulate ON. The particulate ON_p mineralization rate is limited by a phytoplankton concentration dependent Michaelis–Menton relationship, whereas the dissolved ON_d mineralizes directly to NH_4. The NH_4 is taken up by the phytoplankton in proportion to their growth rate and as controlled by the preference factor. The decay of NH_4 to NO_3 is rate limited by the DO concentration. Ammonium can be resupplied from the sediment.

The preference factor is formulated from Thomann and Fitzpatrick (1982) in Equation 13.1.2 in Table 13-2. It is a relatively complex function that goes to zero when the NH_4 concentration decreases to zero and is near unity when NH_4 is large relative to NO_3. Basically, the function states that phytoplankton prefer NH_4 when it is present, but will use NO_3 when NH_4 is not present.

Ammonium can be released from the sediment as indicated by the SED_{nh4} term. Its proportionality to SOD using Spnh3 is based on the relationships developed by Pamatmat (1971). Alternatively, the Spnh3 can be set to zero, and SEDnh3m specified as a uniform value throughout the water body.

13.2.2 Nitrate Nitrogen

The rate relationship for nitrate nitrogen is given in Equation 13.2.1 in Table 13-2. It shows the NO_3 supplied by the decay of NH_4, with the rate being controlled by a DO limitation. The NO_3 is taken up by phytoplankton in proportion to their growth rate and the quantity $(1-P_{NH3})$, demonstrating that the NO_3 is underused when NH_4 is present.

The NO_3 can denitrify. Denitrification takes place at very low DO concentrations and represents a loss of NO_3 from the cycle.

13.2.3 Inorganic Phosphorous

The rate relationship for inorganic phosphorous is given in Equation 13.3.1 in Table 13-2. It is supplied by the respiration of phytoplankton and the mineralization of OP_d and of OP_p as limited by the presence or absence of phytoplankton. The inorganic phosphorous is taken up by the growth of phytoplankton and can be supplied from the sediment.

13.2.4 Phytoplankton

The rate relationship for the growth or decay of phytoplankton is given as Equation 13.4.1 in Table 13-2. It shows that phytoplankton grow or decay in proportion to their own density depending on the net sum of the growth rate, respiration rate, and death rate. The zooplankton-grazing rate is also proportional to the phytoplankton density. Phytoplankton settle out of the water column.

The actual growth rate, G_p, is the maximum growth rate as modified by the temperature limitation, a light-limiting function, and a nutrient-limiting function. The latter takes the minimum value of a function dependent on the presence of nitrogen or dependent on the presence of inorganic phosphorous. Whichever function has the minimum value is considered to be the limiting nutrient.

The light-limiting function depends on the light saturation value, Is, for a particular phytoplankton species and the short-wave solar radiation intensity that decreases through the water column according to Equation 13.4.3. The light intensity is attenuated through the water column by the sum of a constant attenuation coefficient and a term proportional to the phytoplankton density. Phytoplankton can become "self shading" and grow to such a density that they shut off their own light source, reducing the growth rate to a very low value.

The respiration rate is temperature dependent, but the death rate is not. Zooplankton grazing is proportional to the phytoplankton density, which basically states that the more food available for zooplankton, the more they will grow and consume.

13.2.5 Carbonaceous Biochemical Oxygen Demand

The rate relationship for dissolved CBOD_d is given in Equation 13.5.1 in Table 13-2. It shows that CBOD_d increases in proportion to the death rate of phytoplankton and the amount of dissolved carbon contained within the phytoplankton cells. The CBOD_d also increases with the excretion of carbon through the cell walls. The dissolved CBOD_d decays or oxidizes to a form of CO_2 when DO is present. The dissolved CBOD_d can denitrify in the absence of DO and the presence of NO_3.

The particulate form of CBOD_p increases in proportion to the phytoplankton death rate and the amount of particulate organic carbon contained within the phytoplankton cells. It increases from the excretion of the zooplankton. The latter is limited by the assimilation efficiency of zooplankton, AS, which states that the greater the ability of the zooplankton to retain particulate carbon, the less that is released to CBOD_p. The particulate CBOD_p oxidizes to some form of CO_2 similar to the dissolved form of CBOD_d. Particulate CBOD_p settles out.

13.2.6 Dissolved Oxygen

The DO balance is given in Equation 13.6.1 in Table 13-2. The first term is reaeration at the water surface where the DO tends to be driven to its saturation value at the surface temperature of water. Next is uptake of oxygen by CBOD_d and CBOD_p as they oxidize to simpler carbon products, and its rate can be limited by the DO concentration. Next is nitrification, or the decay of NH_4 to NO_3. Oxygen is supplied by phytoplankton through photosynthesis in proportion to their growth rate, but this process can be limited by too large a preference factor. Next is the loss of DO to phytoplankton by their respiration. Dissolved oxygen is taken up by SOD in the bottom layer.

The wind speed dependent surface reaeration coefficient, Equation 13.6.2, is based on the relationship developed by MacKay (1980). The surface DO concentration is computed as a function of the absolute value of the surface temperature (Tk).

Sediment oxygen demand is formulated from the relationship developed by Pamatmat (1971). The SOD results from the settling of CBOD_p, ON_p, and phytoplankton to the bottom. The coefficients each term is multiplied by have been evaluated from field data. Using the Pamatmat (1971) relationship makes SOD variable throughout the water body depending on the distribution of CBOD_p, ON_p, and phytoplankton. In addition, a spatially and temporally constant SODm can be added to the relationship, or it can be used independently of the settling components.

13.2.7 Organic Nitrogen

The rate relationship for ON_d is given in Equation 13.7.1 in Table 13-2. It is produced by the death of phytoplankton in proportion to the fraction of ON_d contained in their cells. It is also excreted through the phytoplankton cell walls. Organic nitrogen decays or mineralizes to NH_4.

The rate relationship for ON_p is given in Equation 13.7.2. It is produced by dying phytoplankton in proportion to the particulate fraction of ON in the cells. The ON_p is also produced from the unassimilated portion of the wastes from zooplankton grazing.

Particulate ON mineralizes to NH_3, and unlike ON_d, its rate is limited by the presence or absence of phytoplankton. Not limiting the mineralization of ON_d allows the WQDPM relationships to function as the ON_d terms in the DOD model and simpler water quality models without the presence of phytoplankton. The particulate component of ON can settle out through the water column.

13.2.8 Organic Phosphorous

The OP relationships are similar to the ON relationships. The rate relationship for OP_d is given in Equation 13.8.1 in Table 13-2. It is produced by the death of phytoplankton in proportion to the fraction of OP_d contained in their cells. It is also excreted through the phytoplankton cell walls. Organic phosphorus decays or mineralizes to PO_4.

The rate relationship for OP_p is given in Equation 13.8.2. It is produced by dying phytoplankton in proportion to the particulate fraction of OP in the cells. The OP_p is also produced from the unassimilated portion of the wastes from zooplankton grazing.

Particulate OP mineralizes to PO_4 and, unlike OP_d, its rate is limited by the presence or absence of phytoplankton. Not limiting the mineralization of OP_d allows the WQDPM relationships to function as the ON_d terms in the DOD model and simpler water quality models without the presence of phytoplankton. The particulate component of OP can settle out through the water column.

13.3   Rates and Constants in the Water Quality Dissolved Particulate Model

Certain rates and parameters in the WQDPM can be varied in the INTROGLLVHT _WQM.dat file for a project using the WQDPM as indicated previously in Table 3-6. The rest of the rates and parameters used in the INTROGLLVHT version of the WQDPM are set at default values. The default values used in the INTROGLLVHT version of the WQDPM are given in Table 3-6 along with the range of values that has appeared in the literature. Most of these parameters are the half-saturation constants used in the Michaelis–Menton type of rate-limiting terms, temperature constants, and fractions of different organic carbon, nitrogen, and phosphorous found in phytoplankton and stoichiometric ratios. Most have a very limited range. The rates and parameters that can be specified in the _WQM.dat file are also indicated.

The more complete Generalized Environmental Modeling System for Surfacewaters (GEMSS) modeling system discussed in Chapter 9, which incorporates the WQDPM as one of its available water quality models, gives access to all of the parameters listed in Table 13-3 through a graphical users interface display of the process diagram.

13.4   Higher-Order Water Quality Modeling

The flow diagram for the higher-order carbon-based water quality model WQMCB is given in Figure 13-2. It treats organic carbon in both dissolved and particulate forms as separate state variables. In addition, it and a number of the other constituents are allowed to have "fast" or liable and "slow" or "refractory" decay rates. These constituents are necessary to link the WQMCB to more complex sediment diagenesis exchange models. In addition, the WQMCB model includes the hydrolysis of particulate constituents into solution as dissolved constituents.

The WQMCB is designed to include a number of phytoplankton species and variable growth and death rates through time. An important process included in the WQMCB is the diurnal vertical migration of dinoflagellates as formulated by Boatman (1999) and Boatman and Buchak (1987). Dinoflagellates move upward through the water column during daylight hours to a level near where the light level is at their optimal saturation light level. They sink toward the bottom at night. The net result is that dinoflagellates produce a DO maximum a few meters below the water surface during the day and their respiration removes oxygen from the bottom portion of the water column at night. They are also important in producing the seasonal oxygen distribution in a water body because they graze on the phytoplankton. Comparisons by Boatman (1999) between using the WQMCB without and with the dinoflagellate migration showed that observed DO distributions could not be reproduced without this process regardless of what parameters were adjusted. The WQMCB is included in the GEMSS modeling system discussed in Chapter 9.

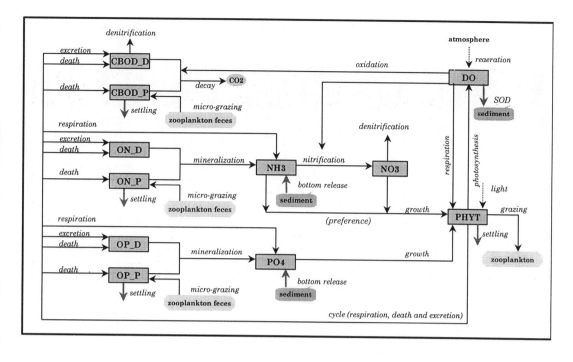

Figure 13-1. Process flow diagram for the WQDPM water quality model.

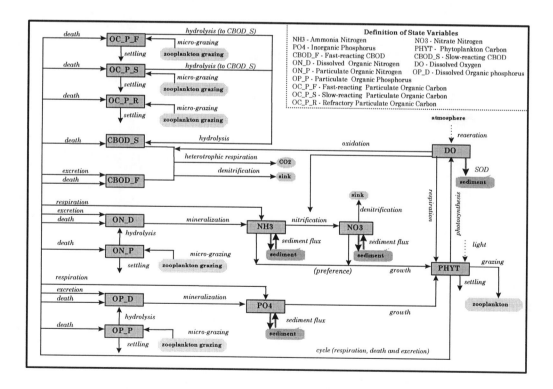

Figure 13-2. Flow diagram for the carbon based water quality model WQMCB.

Table 13-1. Water quality dissolved particulate model index for each constituent and constituent symbol used in equations given in Table 13-2.

WQDPM Index	Symbol Equations	Variable	Notation	Unit
4	CD5	Dissolved carbonaceous BOD	CBOD_D	g O_2/ m^3
5	CD7	Dissolved organic nitrogen	ON_D	g N/ m^3
6	C1	Ammonia nitrogen	NH3	g N/m^3
7	C6	Dissolved oxygen	DO	g O_2/ m^3
8	C2	Nitrate nitrogen	NO3	g N/ m^3
9	CD8	Dissolved organic phosphorus	OP_D	g P/ m^3
10	C3	Inorganic phosphorus	PO4	g P/ m^3
11	C4	Phytoplankton carbon	PHYT	g C/ m^3
12	CP5	Particulate carbonaceous BOD	CBOD_P	g O_2/ m^3
13	CP7	Particulate organic nitrogen	ON_P	g N/ m^3
14	CP8	Particulate organic phosphorus	OP_P	g P/ m^3
15	Algrt	Net algal growth/death rate	Algrt	Per day

BOD, biochemical oxygen demand.

Table 13-2. Constituent source/sink relationships used in the water quality dissolved particlulate model water quality model.

1. Ammonia nitrogen (NH$_3$)

$$\frac{D C_1}{Dt} = D_R\, a_{nc}\, C_4 + k_{71}\Theta_{71}^{T-20}\, C_{D7} + k_{71}\Theta_{71}^{T-20}\left(\frac{C_4}{K_{mNc}+C_4}\right) C_{P7} - G_P\, a_{nc}\, P_{NH3}\, C_4$$

$$\qquad\qquad respiration \qquad\qquad\qquad mineralization \qquad\qquad\qquad growth$$

$$- k_{12}\Theta_{12}^{T-20}\left(\frac{C_6}{K_{NIT}+C_6}\right) C_1 + SED_{NH3} \qquad\qquad (13.1.1)$$

$$\qquad nitrification \qquad\qquad\qquad sediment\ release$$

ammonia preference factor

$$P_{NH3} = C_1\left(\frac{C_2}{(K_{mN}+C_1)(K_{mN}+C_2)}\right) + C_1\left(\frac{K_{mN}}{(C_1+C_2)(K_{mN}+C_2)}\right) \qquad (13.1.2)$$

sediment release of NH

$$SED_{NH3} = Snh3 * SOD \qquad\qquad (13.1.3)$$

2. Nitrate nitrogen (NO$_3$)

$$\frac{D C_2}{Dt} = k_{12}\Theta_{12}^{T-20}\left(\frac{C_6}{K_{NIT}+C_6}\right) C_1 - G_P\, a_{nc}\,(1 - P_{NH3})\, C_4$$

$$\qquad\qquad nitrification \qquad\qquad\qquad growth$$

$$- k_{2D}\Theta_{2D}^{T-20}\left(\frac{K_{NO3}}{K_{NO3}+C_6}\right) C_2 \qquad\qquad (13.2.1)$$

$$denitrification$$

3. Inorganic phosphorus (PO$_4$)

$$\frac{D C_3}{Dt} = D_R\, a_{pc}\, C_4 + k_{83}\Theta_{83}^{T-20}\, C_{D8} + k_{83}\Theta_{83}^{T-20}\left(\frac{C_4}{K_{mPc}+C_4}\right) C_{P8}$$

$$\qquad\qquad respiration \qquad\qquad\qquad mineralization$$

$$- G_P a_{pc} C_4 + SED_{PO4} \qquad (13.3.1)$$

growth   sediment release

4. Phytoplankton (PHYT)

$$\frac{DC_4}{Dt} = ((1 - f_e) G_p - D_R - D_D) C_4 - g_{graze} - \frac{\partial v_{s4} C_4}{\partial z} \qquad (13.4.1)$$

growth-excretion respiration death grazing   settling

growth

$$G_p = k_{1c} \Theta_{1c}^{T-20} \left(\frac{I}{I_s} e^{1 - \frac{I}{I_s}} \right) \qquad (13.4.2a)$$

temperature and light limitation

$$\min \left(\frac{C_1 + C_2}{K_{mN} + C_1 + C_2}, \frac{f_{D3} C_3}{K_{mP} + f_{D3} C_3} \right) \qquad (13.4.2b)$$

nutrient limitation

light extinction

$$I = I_a(t) e^{-1.0(0.38 + 0.005 SUM_{PHYT}/c2chla)Z} \qquad (13.4.3)$$

respiration

$$D_R = k_{1R}(20°C) \Theta_{1R}^{(T-20)} \qquad (13.4.4)$$

death

$$D_D = k_{1D} \qquad (13.4.5)$$

microzooplankton grazing

$$g_{micrograze} = K_{gmicro} \Theta_{KT}^{T-20} C_4 \qquad (13.4.6)$$

macrozooplankton grazing

$$g_{macrograze} = K_{gmacro} \Theta_{KT}^{T-20} C_4 \qquad (13.4.7)$$

total zooplankton grazing

$$g_{graze} = g_{micrograze} + g_{macrograze} \tag{13.4.8}$$

5. Carbonaceous biochemical oxygen demand (CBOD)

$$\frac{DC_{D5}}{Dt} = a_{oc}f_{oc}D_D C_4 + a_{oc}f_e G_p C_4 - k_d \Theta_d^{(T-20)} \left(\frac{C_6}{K_{BOD} + C_6} \right) C_{D5}$$

       death   *excretion   oxidation*

$$-\frac{5}{4}\frac{32}{14} k_{2D} \Theta_{2D}^{(T-20)} \left(\frac{K_{NO3}}{K_{NO3} + C_6} \right) C_2 \tag{13.5.1}$$

denitrification

$$\frac{DC_{P5}}{Dt} = a_{oc}(1 - f_{oc})D_D C_4 + a_{oc}(1 - AS)g_{micrograze}$$

            death   *grazing*

$$- k_d \Theta_d^{(T-20)} \left(\frac{C_6}{K_{BOD} + C_6} \right) C_{P5} - \frac{\partial v_{s5} C_{P5}}{\partial Z} \tag{13.5.2}$$

   oxidation   *settling*

6. Dissolved oxygen (DO)

$$\frac{DC_6}{Dt} = k_2(C_s - C_6) - k_d \Theta_d^{T-20} \left(\frac{C_6}{K_{BOD} + C_6} \right)(C_{D5} + C_{P5}) - \frac{64}{14} k_{12} \Theta_{12}^{T-20} \left(\frac{C_6}{K_{NIT} + C_6} \right) C_1$$

  reaeration (surface)   *heterotrophic respiration*   *nitrification*

$$+ G_P \left(\frac{32}{12} + \frac{48}{14}\frac{14}{12}(1 - P_{NH3}) \right) C_4 - \frac{32}{12}D_R C_4 - SOD \Theta_s^{T-20} \tag{13.6.1}$$

   phyto. growth   *phyto. respiration*   *SOD (bottom)*

DO reaeration coefficient

$$k_2 = 6.416 \cdot 10^{-7}(U_{10})^{1.5} \tag{13.6.2}$$

DO saturation concentration

$$\ln C_s = -139.34 + (1.5757 \cdot 10^5)T_K^{-1} - (6.6423 \cdot 10^7)T_K^{-2}$$
$$+ (1.2438 \cdot 10^{10})T_K^{-3} - (8.6219 \cdot 10^{11})T_K^{-4} \qquad (13.6.3)$$
$$- 0.5535S(0.031929 - 19.428\,T_K^{-1} + 3867.3\,T_K^{-2})$$

sediment oxygen demand

$$SOD = [S_{p5}\,v_{s5}\,C_5 + S_{p7}\,v_{s7}\,C_7 + S_{p4}\,a_{oc}\,v_{s4}\,C_4] + SODm \qquad (13.6.4)$$

7. Organic nitrogen (ON)

$$\frac{DC_{D7}}{Dt} = D_D\,a_{nc}\,f_{on}\,C_4 + a_{nc}\,f_e\,G_p\,C_4 - k_{71}\Theta_{71}^{T-20}\,C_{D7} \qquad (13.7.1)$$

$$\quad\quad\;\; death \quad\quad\quad\quad excretion \quad\quad\quad mineralization$$

$$\frac{DC_{P7}}{Dt} = D_D\,a_{nc}(1 - f_{on})C_4 + a_{ncp}(1 - AS)g_{micrograze}$$

$$\quad\quad\;\; death \quad\quad\quad\quad\quad\quad grazing$$

$$- k_{71}\Theta_{71}^{T-20}\left(\frac{C_4}{K_{mNc} + C_4}\right)C_{P7} - \frac{\partial v_{s7}C_{P7}}{\partial Z} \qquad (13.7.2)$$

$$\quad mineralization \quad\quad\quad\quad settling$$

8. Organic phosphorus (OP)

$$\frac{DC_{D8}}{\partial t} = D_D\,a_{pc}\,f_{op}\,C_4 + a_{pc}\,f_e\,G_p\,C_4 - k_{83}\Theta_{83}^{T-20}\,C_{D8} \qquad (13.8.1)$$

$$\quad\quad\;\; death \quad\quad\quad\quad excretion \quad mineralization$$

$$\frac{DC_{P8}}{Dt} = D_D\,a_{pc}(1 - f_{op})C_4 + a_{pcp}(1 - AS)g_{micrograze}$$

$$\quad\quad\;\; death \quad\quad\quad\quad\quad\quad grazing$$

$$- k_{83}\Theta_{83}^{T-20}\left(\frac{C_4}{K_{mPc} + C_4}\right)C_{P8} - \frac{\partial v_{s8}C_{P8}}{\partial Z} \qquad (13.8.2)$$

$$\quad mineralization \quad\quad\quad\quad settling$$

Table 13-3. Parameters used in water quality dissolved particulate model water quality model equations given in Table 13-2, arranged alphabetically.

Parameter	Default	Range	Definition	Units	Reference
Θ_{2D}	1.08	1.08	Temperature coefficient	None	Ambrose (1993)
Θ_{12}	1.08	1.08	Temperature coefficient	None	Ambrose (1993)
Θ_{1c}	1.068	1.068	Temperature coefficient	None	Ambrose (1993)
Θ_{1R}	1.045	1.045	Temperature coefficient	None	Ambrose (1993)
Θ_{1R}	1.045	1.045	Temperature coefficient	None	Ambrose (1993)
Θ_{2D}	1.045	1.045	Temperature coefficient	None	Ambrose (1993)
Θ_{71}	1.08	1.08	Temperature coefficient	None	Ambrose (1993)
Θ_{83}	1.08	1.08	Temperature coefficient	None	Ambrose (1993)
Θ_{d}	1.047	1.047	Temperature coefficient	None	Ambrose (1993)
Θ_{KT}	1.045	1.045	Temperature coefficient	None	Ambrose (1993)
Θ_{s}	1.08	1.08	Temperature coefficient	None	Ambrose (1993)
a_{nc}	0.25	0.25	Nitrogen to carbon ratio	g N/g C	Ambrose (1993)
a_{ncp}	0.25	0.25	Particulate nitrogen to carbon ratio	None	Ambrose (1993)
a_{oc}	32/12	32/12	Oxygen uptake per unit of algae respired	g O_2/g C	Ambrose (1993)
a_{oc}	32/12	32/12	Oxygen to carbon ratio	g O_2/g C	Ambrose (1993)
a_{pc}	0.025	0.025	Phosphorus to carbon ratio	g P/g C	Ambrose (1993)
a_{pcp}	0.75	0.70 ~ 0.80	Particulate organic phosphorus to carbon ratio	None	Ambrose (1993)
as	0.5	0.5 - 0.8	Assimilation efficiency of zooplankton grazing	None	EPA (1985)
c2chla	30	10 - 100	Ratio of carbon to chlorophyll a	None	EPA (1985)
C_s	Equation (6.3)	Equation (6.3)	Dissolved oxygen saturation	gO_2/m^3	Ambrose (1993)
f_e Specify	0.1	0.1~0.8	Excretion fraction of phytoplankton	None	EPA (1985)
f_{oc}	0.5	0.5	Organic carbon from dead algae	None	Ambrose (1993)
f_{on}	0.5	0.5	Organic nitrogen from dead algae	None	Ambrose (1993)
f_{op}	0.5	0.5	Organic phosphorus from dead algae	None	Ambrose (1993)
I_s Specify	127	110-150	Saturating light intensity	watt/m^2	WDOE (1984)
Gzoo	1.0	0.0-2.5	Multiplier on graze rate	ND	
k_{12} Specify	0.02	(0.09~0.13)	Nitrification rate	day^{-1}	WDOE (1984)
k_{1c} Specify	2.0	2.0	Maximum growth rate	day^{-1}	Ambrose (1993)
k_{1D} Specify	0.015	0.2	Death rate	day^{-1}	WDOE (1984)
k_{1R} Specify	0.125	0.125	Phytoplankton respiration rate, 20°C	day^{-1}	Ambrose (1993)
k_{1R} Specify	0.015	(.05-.20)	Endogenous respiration rate at 20°C	day^{-1}	WDOE (1984)
k_2		Equation (13.6.2)	Reaeration rate @ 20°C	day^{-1}	Ambrose (1993)
k_{2D} Specify	0.09	0.09	Denitrification rate at 20°C	day^{-1}	Ambrose (1993)
k_{71} Specify	.0075	0.01-0.15	Organic nitrogen mineralization rate	day^{-1}	WDOE (1984)
k_{83} Specify	0.22	0.10-0.30	Dissolved organic phosphorus mineralization at 20°C	day^{-1}	Ambrose (1993)
K_{BOD}	0.5	0.5	Half-saturation constant for oxygen limitation	g O_2/m^3	Ambrose (1993)
k_d Specify	0.15	(0.02-0.20)	Deoxygenation rate @ 20°C	day^{-1}	WDOE (1984)

Parameter	Default	Range	Definition	Units	Reference
Kg_{macro}	0.101	0.101	Grazing rate due to macrozooplankton	day^{-1}	Boatman (1997)
Kg_{micro}	0.081	0.081	Grazing rate due to microzooplankton	day^{-1}	Boatman (1997)
K_{mN}	10/1000	25/1000	Half-saturation constant for nitrogen	$g\ N/\ m^3$	WDOE (1984)
K_{mNc}	1.0	1.0	Half-saturation constant for nitrogen mineralization	$g\ C/\ m^3$	Boatman (1997)
K_{mP}	1.0/1000	1.0/1000	Half-saturation constant for phosphorus	$g\ P/\ m^3$	Ambrose (1993)
K_{mPc}	5.0	1.- 10.0	Half-saturation constant for phosphorus mineralization	$g\ C/\ m^3$	Boatman (1997)
K_{NIT}	2.0	2.0	Half-saturation constant for oxygen limitation of nitrification	$g\ 0_2/\ m^3$	Ambrose (1993)
K_{NO3}	0.1	0.1	Michaelis constant for denitrification	$g\ O_2/\ m^3$	Ambrose (1993)
P_{NH3}	Eq. (13-6.2)		Preference for ammonia uptake term	None	Ambrose (1993)
S4 Specify	0.02	0.01-0.09	Algae contribution to sediment oxygen demand	None	Pamamat (1971)
SED_{NH3}	Formula	Data	Sediment release of ammonia	$g\ N/m^2/d$	WDOE (1984)
SED_{NH3m} Specify	0.0	0.25-2.50	Measured. sediment release of NH3	$g\ N/\ m^2/d$	Measurement
SED_{PO4m} Specify	1.5	Field data	Measured sediment release of phosphorus	$g\ P/m^2/d$	Measurement
SODm Specify	0.0	0.5-5.0	Background sediment oxygen release	$gO_2/m^2/d$	WDOE (1984)
Sp5 Specify	0.07	0.01-0.09	CBOD_P contribution to sediment oxygen demand	None	Pamamat(1971)
Sp7 Specify	0.07	0.01-0.09	ON_P contribution to sediment oxygen demand	None	Pamamat(1971)
SpNH3	0.45	0.20-0.60	Fraction of sediment oxygen demand released as NH3		Panamat(1971)
v_{s4} Specify	0.09	(0.05~0.5)	Settling velocity Phytoplankton .	m/day	WDOE (1984)
v_{s5} Specify	0.08	(0.05 0.2)	Settling velocity of CBOD_P	m/day	EPA (1985)
v_{s7} Specify	0.08	(0.05~0.5)	Settling velocity of ON_P	m/day	EPA (1985)
v_{s8} Specify	0.08	(0.05~0.2)	Settling velocity OP_P	m/day	EPA (1985)
Wad Specify	Field data	Field data	Wind speed	m/s	Measurement

Specify ➔ See key parameters in Water Quality Dissolved Particulate Model (WQDMPM) Water Quality Model, Chapter 3, Table 3-5.

CBOD_P, particulate carbonaceous biochemical oxygen demand; NH_3, ammonia; ON_P, particulate organic nitrogen; OP_P, particulate organic phosphorous.

14. SEDIMENT SCOUR AND DEPOSITION RELATIONSHIPS

In this chapter, the detailed sediment transport and scour relationships will first be presented, and then the sediment rate will be defined. Also given is the approximate relationship between the model input sediment parameters.

14.1   Sediment Transport Relationships

The sediment transport relationship is as follows:

$$DC/Dt = -\partial V_s C/\partial z + h_b \tag{14.1}$$

where C is the suspended sediment concentration in g/m^3, V_s is the settling velocity in m/s, and h_b is the bottom scour load in $g/m^2/s$.

The settling velocity is computed from Stokes law as

$$V_s = g(Spgr - 1)d^2/(18\nu) \tag{14.2}$$

where g is the gravitational acceleration, 9.78 m/s^2; Spgr is the specific gravity of the sediment; d is the diameter of the sediment particle in meters; and ν is the molecular viscosity of water in m^2/s.

The molecular viscosity of water can be computed as a function of temperature as follows:

$$\nu = 1.79 \times 10^{-6} \, Exp(-0.0266Temp) \tag{14.3}$$

where Temp is the water temperature in degrees C.

The bottom scour rate, h_b, is computed using the Van Rijn formula given in Nielson (1992, Equation 5.3.2) as

$$h_b = Vscour*Cbottom \tag{14.4}$$

where:

$$Vscour = 0.00033 \, (\theta'/\theta_c - 1)(Spgr - 1)^{0.6} g^{0.6} d^{0.8}/\nu^{0.2} \tag{14.5}$$

where θ_c is the critical Shields parameter and θ' is the bottom friction parameter.

The bottom friction parameter is computed from the relationship of (Nielson 1992, Equation 2.2.4):

$$\theta' = \tau'/[g(Sgr - 1)d] \tag{14.6}$$

where:

$$\tau' = gU_b^2/Ch^2 \qquad\qquad (14.7)$$

is the bottom shear stress of the flow on the sediment and U_b is bottom velocity in m/s and Ch is the Chezy bottom friction in $m^{1/2}/s$.

The Cbottom is computed by applying the balance between the downward advection of the particle with the upward turbulent mixing at the bottom from the relationship of

$$V_s C - Az\partial C/\partial z = 0. \qquad\qquad (14.8)$$

This gives the bottom concentration of

$$Cbottom = C(dz)\ Exp(Vs\ dz/Dz) \qquad\qquad (14.9)$$

where C(dz) is the finite difference model concentration at a distance of dz off the bottom, dz is the finite difference model bottom layer thickness, Vs is the settling velocity in m/s, and Dz is the vertical turbulent dispersion coefficient in the bottom layer.

For scour to take place, there has to be some bottom sediment concentration. Hence, Cbottom is computed as the maximum of the value given by Equation 14.9 or the Cbotmin specified in the input data. Estimates of Cbotmin are given in Table 14-1 as a function of particle diameter derived from the tidal channel model simulations using data provided by Schubel et al. (1978).

14.2   Definition of Sediment Rate

The sediment rate is defined as the long-term average of the difference between the settling rate and the scouring rate. The formulation is

$$Sedrate = 1/t_s\ \Sigma(V_s C(dz) - VscourCbottom)dt \qquad\qquad (14.10)$$

where dt is the computational time step in seconds and t_s is the length of simulation.

The sediment rate is usually converted to units of $g/m^2/yr$. If the Sedrate is positive, then bottom sediment is building up over time; if it is negative, then there is net scouring.

The instantaneous sediment rate, Isdrte, is defined as follows:

$$Isdrte = V_s C(dz) - VscourCbottom. \qquad\qquad (14.11)$$

For steady-state cases, the average sediment rate, Sedrate, will become equal to the instantaneous sediment rate. For tidal and time-varying cases, the Isdrte will describe the variation around Sedrate.

14.3   Relationship Between Model Input Parameters

The input parameters for the sediment model shown in Chapter 3, Section 3.3.4 are the sediment particle diameter, diased; the sediment specific gravity, Spgr; the critical Shields parameter, Shldp; and the minimum bottom sediment concentration, cbotmin. Very approximately, the larger the diased, then the larger the Spgr and the Cbotmin as shown in Table 14-1. The Shields parameter ranges from 0.03 to 0.08 (Nielson 1992, Figure 2.2.2 and Equation 2.2.7), and depends on the cohesivness of the bottom sediments.

Table 14-1. Approximate relationships between sedimentation parameters.

Diased, mm	0.02	0.20	2.0
Spgr	1.2	1.8	2.2
Cbotmin, mg/l	1.0	3.0	5.0

GLOSSARY

baroclinic flow. Circulation induced by horizontal density differences.

barotropic flow. Circulation induced by differences in surface elevation.

bottom scour load. The rate of sediment transfer from the bottom of a water body into the water column.

boundary conditions. The external conditions driving the circulation within a water body including inflows, outflows, tides, and winds.

brine. Water with a very high salt concentration.

carbon-based model. A water quality model that includes organic carbon as well as organic nitrogen and organic phosphorous as a state variable.

Chezy friction coefficient. The coefficient relating bottom velocity to bottom shear.

circulation. The three-dimensional flow field within a water body.

coefficient of surface heat exchange. The coefficient describing the rate at which heat is exchanged across the water surface.

computational time step. The increment of time at which the numerical computation is proceeding.

constituent relationship. An equation that describes how fast a constituent concentration is changing due to different biochemical processes.

Coriolis acceleration. An artificial acceleration imposed on the fluid due to the fixed coordinate system rotating with the earth about its axis.

curtain wall. An interior barrier or boundary within a water body that forces the circulation to go over, under, or around it.

decay rate. The rate at which a constituent decreases in concentration due to a biochemical or radioactive process.

dew point temperature. The temperature at which a parcel of air becomes saturated with water vapor.

dinoflagellates. Mobile plankton that undergo vertical migration in response to light intensity.

dissolved oxygen depression. The amount that dissolved oxygen is reduced due to flows into a water body.

diurnal. An event with a daily cycle.

dummy cells. The first or outer row or column of cells in the model grid.

equation of state. The relationship describing water density in terms of temperature, salinity, and pressure.

equilibrium temperature of surface heat exchange. The water surface temperature at which the heat entering a water body due to short-wave solar radiation and atmospheric long-wave radiation equals the loss of heat from the water due to long-wave back radiation, evaporation, and conduction.

eutrophication. Excessively high densities of phytoplankton, usually accompanied by low dissolved oxygen levels due to high nutrient inflows to a water body.

excess temperature. The increment of temperature related to a heat source inflow other than surface heat exchange.

first-order decay. The rate of decrease of a constituent in proportion to its concentration.

flushing time. The time it takes for a simulated initial dye concentration to decrease by one half in a specific model cell.

grazing rate. Rate at which zooplankton consume phytoplankton.

groundwater. Water that enters the water body below the water surface elevation from the surrounding land.

half-life. The time required for the concentration of a constituent to reduce by one half when undergoing decay.

heat rejection rate. The rate at which a cooling water discharge puts heat into a water body.

hydrodynamic modeling. The modeling of the flow field, surface elevations, and at least the temperature and salinity as they affect density, due to the model boundary conditions.

hypolimnion. The portion of a water body that is below the point on a temperature profile where temperature changes most rapidly with depth.

iteration time step. *See* computational time step.

jetty. A wall sticking out into a water body. *See also* curtain wall.

kinetic reaction rate. The rate at which a particular biochemical process proceeds in a water quality model.

macrozooplankton. Large zooplankton.

microzooplankton. Small zooplankton.

mineralization. The process by which organic compounds are transformed into ionic compounds.

model grid. The overall rectangular grid on which the water depths are placed.

model grid cell. The surface dx by dy cell on the model grid, and sometimes the whole dx by dy by dz cell volume.

nutrient cycle. A description of the processes by which nutrients are passed into and out of phytoplankton.

particulate-based model. The WQDP model that includes both the particulate as well as the dissolved form of organic nitrogen and organic phosphorous.

photosynthesis. The process by which phytoplankton produce dissolved oxygen using light as an energy source.

phytoplankton. Small, immobile organisms that drift with currents and can produce dissolved oxygen due to photosynthesis.

plant heat rejection. *See* heat rejection rate.

pumping rate. The rate at which water is pumped through a facility such as a power plant.

rate of decay. *See* decay rate.

reaeration. The coefficient describing the rate at which dissolved oxygen is exchanged across the water surface.

residence time. *See* flushing time.

respiration. The uptake of dissolved oxygen by phytoplankton in the absence of sunlight.

response temperature. The temperature in a mixed column of water that results due to surface heat exchange alone.

response time. The time required for the response temperature to reach within a certain limit of the equilibrium temperature.

Richardson number. The ratio of the vertical density gradient to the velocity shear.

salinity stratification. The variation of salinity with depth.

sediment rate. The product of the settling velocity and the concentration near the bottom of sediment.

semi-implicit solution. The combined solution for the surface elevations and velocity distributions in the hydrodynamic model.

settling velocity. The velocity at which a sediment particle settles to the bottom in quiescent water.

Shield's parameter. The ratio of the bottom shear stress to the square of the settling velocity. It allows one to determine the shear stress at which a sediment particle is resuspended.

skimmer wall. *See* curtain wall.

source/sink terms. The terms describing the rate at which constituents are added to or removed from a water body due to inflows and withdrawals.

Stokes law. The name of the formula for the settling velocity of a sediment particle in terms of the viscosity of water, the specific gravity of the sediment, and the particle diameter.

surface wind stress. The shear stress placed on a water surface due to the wind.

tank test. A modeling configuration that includes no inflows and outflows.

Torrence condition. The condition that applies to numerical forms of the constituent transport relationships that states no more flow can be removed from a cell volume than the volume of the cell itself over a computational time step.

vapor pressure slope. The change in water vapor pressure with water temperature.

Von Karman constant. A constant found from experimental data that allows one to relate the shear within a water column to the horizontal velocity shear.

water quality modeling. The computation of water quality constituent concentrations using formulations that include different biochemical rate processes from the inflows into a water body.

wind shear coefficient. A coefficient that allows one to relate the square of the wind speed to the surface wind shear.

zooplankton. Large, somewhat mobile plankton that graze on phytoplankton and excrete organic nutrients mostly in particulate form.

REFERENCES

Ambrose, R .B., T. A. Wool, and J. L. Martin. 1993. *The water quality analysis simulation program. WASP5. Part A: Model documentation. Part B: The WASP5 model input dataset.* Athens, GA: Environmental Protection Agency (EPA).

Boatman, C. D. 1997. *Field parameter evaluation for Budd Inlet studies.* Seattle, WA: Aura Nova Consultants, Inc.

Boatman, C. D. 1999. *LOTT NPDES permit modifications modeling.* Wayne, PA: Aura Nova Consultants, Inc. and J. E. Edinger Associates, Inc.

Boatman, C. D., and E. M. Buchak. 1987. Application of an ecosystem/water quality model as a tool for managing estuarine water quality. In *Proceedings of the fifth symposium on coastal ocean management,* Vol. 3. Reston, VA: American Society of Civil Engineers (ASCE):3932–3945.

Brown, L. C., and T. O. Barnwell, Jr. 1987. *The enhanced stream water quality model QUAL2E and QUAL2E-UNCAS: Documentation and users manual.* Athens, GA: Environmental Protection Agency (EPA). Report EPA/600/3–87/007.

Buchak, E. M., and J. E. Edinger. 1984. *Generalized, longitudinal-vertical hydrodynamics and transport: Development, programming, and applications.* Wayne, PA: J. E. Edinger Associates, Inc. Prepared for U. S. Army Corps of Engineers Waterways Experiment Station, Vicksburg, Mississippi. Contract No. DACW39–84–M–1636. Document No. 84–18–R.

Buchak, E. M., J. E. Edinger, and V. S. Kolluru. 2001. Simulation of cooling-water discharges from power plants. *J Environ Management* Vol. 61, No 1, Jan. 2001, pp.77-92.

Cameron, W. M, and D.W. Pritchard. 1965. Estuaries. In *The sea*, Vol. II. New York: John Wiley:306–324.

Cerco, F. C., and T. M. Cole. 1993. *Users guide for the CE-QUAL-ICM three-dimensional eutrophication model.* Vicksburg, MS: U. S. Army Corps of Engineers Waterways Experiment Station.

Cole, T. M., and E. M. Buchak. 1995. *CE-QUAL-W2: A two-dimensional, laterally averaged hydrodynamic and water quality model. Version 2.0 users manual.* Vicksburg, MS: U. S. Army Corps of Engineers Waterways Experiment Station. Instruction Report EL–95–1.

Churchill, M. A., H. L. Elmore, and R. A. Buckingham. 1962. The prediction of stream reaeration rates. *ASCE J Sanitary Engrg Division* 88 (SA4):1–46.

Davis, M, and J. Coughlan. 1981. *A model for predicting chlorine concentration within marine cooling circuits and its dissipation at outfalls, in water chlorination environmental impact and health effects.* Vol. 4. Edited by R. L. Jolley et al. Ann Arbor, MI: Ann Arbor Science (The Butterworth Group).

Defant, A. 1958. *Ebb and flow: The tides of earth, air, and water.* Ann Arbor: The University of Michigan Press.

Edinger, J. E., D. K. Brady, and J. C. Geyer. 1974. *Heat exchange and transport in the environment: Electric power research institute cooling water studies.* Baltimore, MD: The Johns Hopkins Univeristy Department of Geography and Environmental Engineering Research. Project RP-49, Report 14.

Edinger, J. E., and E. M. Buchak. 1980. *Numerical hydrodynamics of estuaries.* In Estuarine and Wetland Processes with Emphasis on Modeling edited by P. Hamilton and K. B. Macdonald. New York: Plenum Press:115 –146.

Edinger, J. E., and E. M. Buchak. 1985. Numerical waterbody dynamics and small computers. In *Proceedings of ASCE 1985 hydraulic division specialty conference on hydraulics and hydrology in the small computer age.* Reston, VA: American Society of Civil Engineers.

Edinger, J. E., E. M. Buchak, and M. D. McGurk. 1994. *Analyzing larval distributions using hydrodynamic and transport modeling. Estuarine and coastal modeling III.*: American Society of Civil Engineers, Reston, VA..

Edinger, J. E., and E. M. Buchak. 1995. Numerical intermediate and far field dilution modelling. *J. Water Air Soil Pollution* 83:147–160.

Edinger, J. E., E. M. Buchak, and V. Kolluru. 1998. Flushing and mixing in a deep estuary. *J. Water Air Soil Pollution* 102:345–353.

Edinger, J.E., V. S. Kolluru. 1998. Combined hydrodynamic and water quality modeling for waste water impact studies. In *Proceedings of the mid-Atlantic industrial wastes conference*, edited by L. Christensen. Technomic Publishing Co.,Lancster, PA.

Edinger, J. E., and V. S. Kolluru. 1999. Implementation of vertical acceleration and dispersion terms in an otherwise hydrostatically approximated three-dimensional model. In *ASCE estuarine and coastal modeling: Proceedings of the 6th international conference, November 3–5, 1999, New Orleans, Louisana.* Reston, VA: American Society of Civil Engineers.

Edinger, J. E., and V. S. Kolluru. 2000. Power plant intake entrainment analysis. *ASCE J. Energy Engrg.* 126 (1):pp 1-14.

Edinger, J.E., V.S. Kolluru, and S. Dierks. 2001. Density dependent grazing in estuarine water quality models. Submitted for publication in *Water, Air and Soil Pollution.*

Environmental Protection Agency (EPA) 1971. *Effect of geographical location on cooling pond requirements and performance*. Water Pollution Control Research Series. Washington, DC: Environmental Protection Agency, Water Quality Office. Report No. 16130 FDQ.

Environmental Protection Agency (EPA).1985. *Rates, constants, and kinetics formulations in surface water quality modeling*. 2nd ed. Athens, GA: Environmental Research Laboratory. EPA/800/3-85/040.

Gentleman, W., A. Leising, B. Frost, J. .Murry, and S.Strom. 2000. Dynamics of food-web models with multiple nutritional resources: A critical review of implicit assumptions. JGOFS SMP Workshop. School of Oceanography, University of Washington, Seattle, Washington. July 2000.

Goetchius, K. 2000. Inflow, transport, and fate of atrazine in Upper Chesapeake Bay. Master's thesis, Department of Geography and Environmental Engineering, The Johns Hopkins University.

Howard, R. H., R. S. Boethling, W. F. Jarvis, W. M. Meylan, and E. M. Michalenko. 1991. *Handbook of environmental degradation rates*. : Lewis Publishers, Inc.

Jolley, R. L., and J. H. Carpenter. 1983. A review of the chemistry and environmental fate of reactive oxidant species in chlorinated water. In *Water chlorination: Environmental impact and health effects*. Vol. 4, Book 1: Chemistry and water treatment, edited by R. L. Jolley, W. A. Brungs, J. A. Cotruvo, R. B. Cumming, J. S. Mattice, and V.A. Jacobs. Ann Arbor, MI: Ann Arbor Science (The Butterworth Group).

Kolluru, V. S., E. M. Buchak, and J. E. Edinger. 1998. Integrated model to simulate the transport and fate of mine tailings in deep waters. In *Proceedings of tailings and mine waste 1998*. Balkema Press Rotterdam..

Kolluru, V. S., E. M. Buchak, J. Wu, 1999. Use of Membrane Boundaries to Simulate Fixed and Floating Structures in GLLVHT. In Spaulding, M.L, H.L. Butler (eds.). *Proceedings of the 6th International Conference on Estuarine and Coastal Modeling*. American Society of Civil Engineers, Reston, VA. pp. 485 - 500.

Lauff, G. H, ed. 1967. *Estuaries*. Washington, DC: American Association for the Advancement of Science. Publication No. 83.

Leenderste, J. J, and S.-K. Liu. 1975. *A three-dimensional model for estuaries and coastal seas*. Vol. II, *Aspects of computation*. Santa Monica, CA: Rand Report R-1764-OWRT.

Lung, W. S. 1993. *Water quality modeling*. Vol. 3, *Applications to estuaries*. Boca Raton, FL: CRC.

Mackay, D. 1980. Solubility, partition coefficients, volatility, and evaporation rates. In *Reactions and processes handbook*. Vol. 2, Part A. New York: Springer-Verlag.

Martin, J. L., and S. C. McCutcheon.1998. *Hydrodynamics and transport for water quality modeling*. Boca Raton, FL: Lewis Publishers.

Neumann, G., and W. J. Pierson. 1966. *Principles of physical oceanography*. Englewood Cliffs, NJ: Prentice-Hall:191–196.

Nielson, P. 1992. *Coastal bottom boundary layers and sediment transport*. Vol. 4, *Advanced series on ocean engineering*. River Edge, NJ: World Scientific Publishing Co. Ltd.

Okubo, A. 1971. Oceanic diffusion diagrams. *Deep-Sea Res*. 18:789.

Pamatmat, M. M. 1971. Oxygen consumption by the seabed. IV. Shipboard and laboratory experiments. *Limnology Oceanography* 16:536–550.

Pritchard, D.W. 1952. Estuarine hydrography. *Advances in Geophysics* 1:243–280.

Pritchard, D.W. 1955. Estuarine circulation patterns. *Proc. Am. Soc. Civil Engrg*. 81:

Pritchard, D.W. 1956. The dynamic structure of a coastal plain estuary. *J. Marine Res*. 15:33–42.

Pritchard, D.W. 1967. Observations of circulation in coastal plain estuaries. In *Estuaries*. Washington, DC: American Association for the Advancement of Science. Publication No. 83:37-44.

Pritchard, D.W. 1969. Dispersion and flushing of pollutants in estuaries. *ASCE J. Hyd. Div*. 95 (HY1):115 –124.

Schubel, J. R., R. E. Wilson, and A. Okubo. 1978. Vertical transport of suspended sediment in Upper Chesapeake Bay. In *Estuarine transport processes*, edited by B. Kjerfve. Columbia, SC: University of South Carolina Press:161–176.

Shen, H.H., Cheng, A.H-D., Teng, M.H., Wang, K-H., and Clark, C.K. (Eds.) 2002. Environmental Fluid Mechanics-Theories and Applications. Engineering Mechanics Division/Fluids Committee. ASCE Press, Reston, VA.

Smagorinsky, J.1963. General Circulation Experiments with the Primitive Equations,1,The Basic Experiment, *Monthly Weather Review, Vol 91, pp. 90-164*.

Thomann, R.V., and J. F. Fitzpatrick. 1982. *Calibration and verification of a mathematical model of the eutrophication of the Potomac estuary*. Report by Hydroqual, Inc., Mahwah, NJ, to DES, Washington, DC.

Thomann, R.V., and J. A. Mueller. 1987. Principles of surface water quality modeling and control. New York: Harper & Row.

WDOE (State of Washington Department of Environment). 1984. Comprehensive circulation and water quality study at Budd Inlet, Southern Puget Sound water quality assessment study. Prepared by URS Corporation for the Washington State Department of Ecology.

Wetzel, R.G. 2000. *Limnology*. 3rd ed. Philadelphia: WB Saunders.

Index

Please note: Pages in *italics* refer to figures or tables.